D1607471

WATER BOUNDARIES

WILEY SERIES IN SURVEYING AND BOUNDARY CONTROL

Roy Minnick, Series Editor

Cole/WATER BOUNDARIES

Smith/INTRODUCTION TO GEODESY: THE HISTORY AND CONCEPTS OF MODERN GEODESY

Wolf & Ghilani/ADJUSTMENT COMPUTATIONS: STATISTICS AND LEAST SQUARES IN SURVEYING AND GIS

WATER BOUNDARIES

George M. Cole, P.E., R.L.S.

A WILEY-INTERSCIENCE PUBLICATION

JOHN WILEY & SONS, INC.

New York Chichester Weinheim Toronto Singapore Brisbane

This text is printed on acid-free paper.

Library of Congress Cataloging in Publication Data:

Cole, George.
Water boundaries / George Cole.
p. cm.—(Wiley series in surveying and boundary control)
ISBN 0-471-17929-9 (cloth alk. paper)
1. Boundaries (Estates)—United States. 2. Surveying—Law and legislation—United States. 3. United States—Boundaries.
I. Title. II. Series.
KF639.C65 1997
346.7304'32—dc20
[347.306432] 96-44730

Printed in the United States of America

10 9 8 7 6 5 4 3 2

And as for the western border, ye shall even have the Great Sea for a border; this shall be your west border.

Numbers 34:6

CONTENTS

PREFACE

Water boundaries are perhaps the oldest and most widely used of man's boundaries. Yet, despite this long history of usage, water boundaries are probably, in today's society, the most frequently and bitterly contested boundaries. The edge of water forms an excellent natural boundary in that it is easily defended and easily recognized. However, when attempts are made to precisely locate water boundaries, complex technical and legal quagmires may result. This is primarily because the land/water interface is dynamic. The surfaces of most water bodies are constantly changing due to tides and/or meteorological conditions. In addition, the shoreline in many areas is subject to erosion and accretion caused by waves and currents. Therefore, unlike most other boundaries that are two-dimensional, one must consider a third dimension—height—and a fourth dimension—time—when dealing with water boundaries. Thus, these unique boundaries must be considered in a different manner than conventional land boundaries. Consequently, unique laws and techniques have developed for defining and locating water boundaries.

Recent land use practices have created a growing demand for precisely located and legally defensible water boundaries. This has resulted in a need for literature describing currently accepted theories and techniques for locating such boundaries. Because many aspects of water boundaries are interrelated, background in the entire sphere of water boundaries is helpful for proficiency in any of the areas of specialization within this field. Therefore, comprehensive literature providing coverage of all of the various types of water boundaries is especially needed.

This treatise is an attempt to fulfill that need. It is intended to provide a comprehensive overview of both the legal and technical aspects of the unique and specialized area of water boundaries. Although originally written for surveying students and practicing surveyors, this text also should be helpful and of interest to a wide range of individuals involved with coastal or submerged land. These include attorneys involved with water boundary issues, public land managers, title and real estate professionals, and others dealing with land planning, land development, offshore mineral extraction, and other related fields.

It is noted that there is considerable variation in laws regarding water boundaries from state to state. Therefore, care should be taken to assure applicability when using the information contained herein.

As may be seen from the information presented within this text, some areas within the field of water boundaries have developed to the point of being well defined—both legally and technically. Other areas need further judicial or legislative clarification, as well as technical research. It is hoped this text will serve to stimulate study and clarification of these problem areas as well as provide a perspective on the various aspects of water boundaries. At the conclusion of the text, all literature and case law cited in the text is referenced. For readers interested in further information on a particular aspect, these references offer a starting point.

Sincere appreciation is expressed to those who have contributed to or otherwise assisted in the preparation of this edition.

Tallahassee, Florida GEORGE M. COLE

1

TIDAL SOVEREIGN/ UPLAND WATER BOUNDARIES

1.1 BACKGROUND AND HISTORY

Currently, within the United States, it is generally accepted that the individual states hold title on behalf of the public to most of the submerged lands under navigable waters within their respective boundaries. This ownership is by virtue of what is known as the *public trust doctrine* that, although chiefly a product of English common law, has roots as far back as the ancient Roman Civil Code of Emperor Justinian I written about 500 A.D. Under that code, the sea as well as rivers were considered to be *res communes* or commonly owned by all mankind (Cole 1991).

At the rise of the great maritime nations of the Middle Ages, some nations laid exclusive claim to entire seas or oceans, as opposed to that early Roman concept of a commonly owned resource. Gradually, however, such claims were found to be unrealistic and impossible to defend. Therefore, a new doctrine appeared with claims by various nations limited to a marginal sea of a width that could be defended.

Perhaps the earliest mention of this theory in English literature was by Thomas Digges, an engineer, surveyor, and lawyer during the reign of Queen Elizabeth I, in a book titled *Proofs of the Queen's Interest in Land Left by the Sea and the Salt Shores Thereof.* This treatise formed the basis for the Crown's claim to the submerged lands of the kingdom.

In the United States, this doctrine began to take its present shape with a series of U.S. Supreme Court cases beginning with *Martin v. Waddell* in 1842. According to the Court in that case:

> When the revolution took place the people of each state became themselves sovereign and in that character hold the absolute right to all their navigable waters in the soils under them for their own common use. . . .

In 1845, in *Pollard's Lessee v. Hagan,* the court ruled that states admitted to the Union after the original thirteen colonies also had these rights. Thus, it may be generally stated that the several states, in their sovereign capacity, hold title to the beds under navigable waters. With this background, we may now address the key issue of this discussion. This issue is the limit of sovereign ownership. In this chapter, we are concerned with the inshore limit, that is, where the sovereign submerged lands meet the uplands subject to private ownership.

1.2 BOUNDARY DEFINITIONS IN TIDAL WATERS

The following defines the sovereign/upland boundary in tidally affected waters by examining existing common and statutory law. This examination begins with English case law and continues into U.S. case and statutory law.

Following the claims of submerged lands in behalf of the Crown made by Thomas Digges circa 1568 (see the previous section), there apparently was not immediate judicial acceptance of the claim. In the following century, however, the doctrine became generally accepted, as evidenced by the writings of Lord Matthew Hale, a jurist who was to become the British Chief Justice. He espoused the public trust doctrine in his treatise *De Jure Maris,* written about 1666. In that writing, Hale concluded that the foreshore, which is overflowed by "ordinary tides or neap tides, which happen between the full and change of the moon," belonged to the Crown.

With our knowledge of the tides today, it is obvious that Lord Hale was incorrect in equating "neap tides" with "ordinary tides." At the very least, his definition was ambiguous. In 1854, this definition was clarified in English common law by the case of *Attorney General v. Chambers.* The Chambers case, reflecting tidal theory developed after Hale's writings, ruled that the ordinary high water mark was to be found by "the average of the medium tides in each quarter of a lunar evolution during the year (which line) gives the limit, in the absence of all usage, to the rights of the Crown on the seashore."

In the United States, there apparently was no clarification to the boundary until 1935 and the U.S. Supreme Court's landmark decision in *Borax Consolidated Ltd. v. City of Los Angeles.* In essence, this decision called for application of modern scientific techniques for precisely defining the boundary.

> In view of the definition of the mean high tide, as given by the United States Coast and Geodetic Survey that "mean high water at any place is the average height of all the high waters at that place over a considerable period of time," and the further observation that "from theoretical considerations of an astronomic character" there should be "a periodic variation in the rise of water above sea level having a period of 18.6 years," the Court of Appeals directed that in order to ascertain the mean high tide line with requisite certainty in fixing the bound-

> ary of valuable tidelands, such as those here in question appear to be, "an average of 18.6 years should be determined as near as possible." We find no error in that instruction.

As this language demonstrates, the Borax decision applied modern technical knowledge and set forth a workable technique for precisely locating the boundary in question that still prevails in U.S. common law.

As may be seen from the preceding definition as it evolved, the mean high water line represents an attempt to define the upper reach of the daily tide as the boundary between publicly owned submerged lands and uplands subject to private ownership. Because the upper reach of the tide varies from day to day, the use of an average values attempts to split the difference for a compromise line. This results in a line that is exceeded by the high tide on approximately one-half of the tidal cycles.

Case law in the various coastal states has, in the main, followed the English common law and its updated definition as put forth in the Borax decision. Sixteen states (Alabama, Alaska, California, Connecticut, Florida, Georgia, Maryland, Mississippi, New Jersey, New York, North Carolina, Oregon, Rhode Island, South Carolina, Texas, and Washington) have followed this course (Maloney and Ausness 1974; Cole 1977). Some states have codified their common law on this subject. For example, in Florida, the Coastal Mapping Act of 1974 (Chapter 177, Part II, Florida Statutes) declares that "mean high water line along the shores of lands immediately bordering on navigable waters is recognized and declared to be the boundary between the foreshore owned by the State in its sovereign capacity and upland subject to private ownership." The Statute also defines the mean high water line using the Borax definition.

It should be noted that there are exceptions to these generalized statements. For example, six Atlantic Coast states (Delaware, Maine, Massachusetts, New Hampshire, Pennsylvania, and Virginia) recognize the mean *low* water line as the sovereign/upland boundary. For many of these low water states, that boundary is based on an early Massachusetts colonial ordinance of 1641–1647 that provided as follows:

> . . . in all creeks, coves and other places, about and upon salt water where the Sea ebs and flows, the Proprietor of the land adjoining shall have proprietie to the low water mark where the Sea doth not ebb above a hundred rods, and not more wheresoever it ebs farther.

In addition, exceptions exist where civil law, as opposed to Anglo/American common law, controls. This is generally in areas where the land title has its roots in a grant from a sovereign power where civil law, such as the Roman Institutes of Justinian, prevailed. Possibly because the Roman civil law code was developed in an area with minimal daily tidal range (the Mediterranean Sea), it does not define the coastal boundary in terms of daily tide, but rather in terms of seasonal water level changes. As an example, a translation of a portion of that code reads as follows:

> The sea-shore, that is, the shore as far as the waves go at furthest, was considered to belong to all men. . . . The sea shore extends as far as the greatest winter floods runs up. (Sanders 1876)

The difference between the common law and civil law boundary definitions was clearly made in the previously discussed Borax decision as follows:

> By the civil law, the shore extends as far as the highest waves reach in winter. But by the common law, the shore "is confined to the flux and reflux of the sea at ordinary tides."

As an example of the use of the civil law definition, Louisiana has adopted the civil law boundary of the line of the highest winter tide. The State of Hawaii also appears to follow this tradition with a coastal boundary defined as "the upper reach of the wash of the waves" (Maloney and Ausness 1974). Another such example is the State of Texas, which has recognized the civil law definition in a number of cases in areas of the state with origins of land title in Spanish or Mexican land grants. For such grants, it has been held that the limit of ownership is controlled by old Spanish law contained in *Las Sieta*

Partidas, written in the thirteenth century, that closely tracts the Roman Institutes of Justinian.

In 1958, the Supreme Court of Texas *(Luttes v. State)* attempted to apply the use of daily tides to civil law definitions. In that trial, testimony provided various interpretive translations of *Las Siete Partidas* that suggest that the proper meaning is alternatively the highest swell of the year, the highest tide of the year, or an average high tide. The Court concluded that the language of *Las Siete Partidas* implied an average tide and that the applicable rule is that of the "average of highest daily water computed over or corrected to the regular tidal cycle of 18.6 years." In a response to a motion for rehearing, the court clarified the intent as follow:

> It was our intention to hold, and we do hold, that the line under the Spanish (Mexican) law is that of mean higher high tide, as distinguished from the mean high tide of the Anglo-American law.

Regarding this decision, it is noted that some coastal areas in Texas have a relatively small average range of tide in relation to the annual variation related to the sun's declination and seasonal winds. In some such areas, the relatively large seasonal variation, combined with an extremely flat coastal slope, results in wide tidal flats that may be either exposed or inundated, depending on the season of the year and wind conditions. In such areas, the line of mean higher high water may be found considerably waterward of that Spanish grant shoreline indicated by the grant confirmation surveys. Therefore, mean higher high water, although offering mathematical certainty, may not represent the true intent of such grants in all areas. Recently, a trial court decision in Texas *(Bright et al. v. Mauro et al.)* held that the methodology suggested in the Luttes decision would not apply in such areas. In that case, the court accepted a boundary determined by physical shoreline features located many miles landward of the mean higher high water line using daily tidal signals. Another possible alternative for such areas is discussed in Section 1.3.5 of this text.

1.3 TECHNIQUES FOR LOCATING TIDAL BOUNDARIES

1.3.1 Tidal Constituents

The tide is the alternating rise and fall in sea level produced by the gravitational force of the moon and the sun. Other non-astronomical factors, such as meteorological forces, ocean floor topography, and coast line configuration, also play an important role in shaping the tide (National Ocean Survey 1976). To understand the mechanics of the tide producing forces, one may visualize a moon orbiting around an earth covered only with a layer of water, as illustrated in Figure 1.1 (Zetler 1959). Due to the attractive force of the moon, there would a bulge in the water on both sides of the earth in line with the moon and a low water zone in between. The high water on the side of the earth closest to the moon is caused by the moon's pull on the fluid water. On the side opposite the moon, the lesser gravitational force and the centrifugal force as the earth

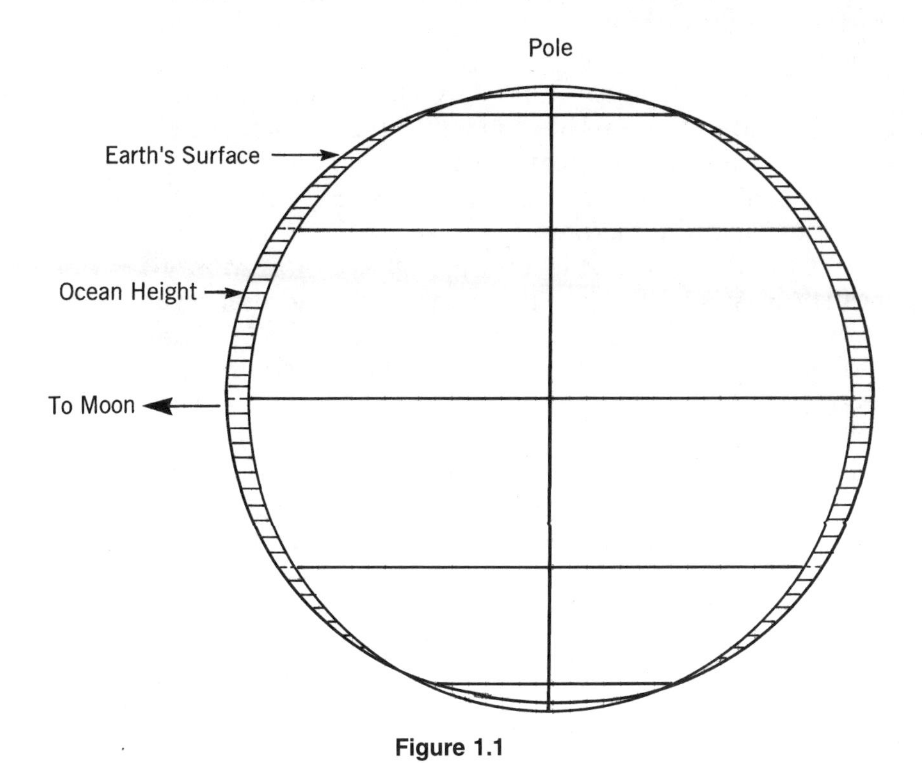

Figure 1.1

and moon spin causes the high water. These bulges of water follow the moon in its revolution about the earth.

Because the average interval between consecutive transits of the moon (upper and lower) is 12.42 hours, the moving high waters take the form of a sine wave with a period of 12.42 hours, as depicted in Figure 1.2. Similarly, there is a sine wave with a period of 12.00 hours following the apparent rotation of the sun.

The sea-level rises caused by many other relationships among the sun, earth, and moon may also be considered as sine waves of a specific periods. For example, the elliptical orbit of the moon about the earth results in a constituent wave with a period of 27.55 days with the highest water at the time of perigee (when the moon is closest to the earth) and the lowest water when the moon is the greatest distance away. In addition to the constituents of different periods, another sinusoidal cycle, that associated with the regression of the moon's nodes[1] with a period of 18.6 years, affects the amplitudes of the various constituents. The resultant tide is the composite, or algebraic sum, of all the previously mentioned constituent cycles as modified by the 18.6-year nodal cycle.

Although there are theoretically several hundred tidal constituents, not all are significant. The 37 constituents considered significant for tidal predictions by the National Ocean Service are provided in Table 1.1.

It is noteworthy that when the high water of more than one constituent is in phase, tides higher than normal occur. Such is the case twice a month when the moon's and the sun's principal constituents are in phase. This occurs near the time of the new and full moon when the earth, moon, and sun are in a line and produce the so-called *spring tides*. *Neap tides* are those that occur at the time of the quarter or three-quarter moon when the sun and moon are at 90 degrees to each other as measured from the earth. Their respective following waves are then out of phase and result in lower high waters.

Figures 1.3 and 1.4 illustrate some of the major constituent

[1] Regression of the moon's nodes refers to the movement of the intersection of the moon's orbital plane and the plane of the earth's equator, which completes a 360-degree circuit in 18.6 years.

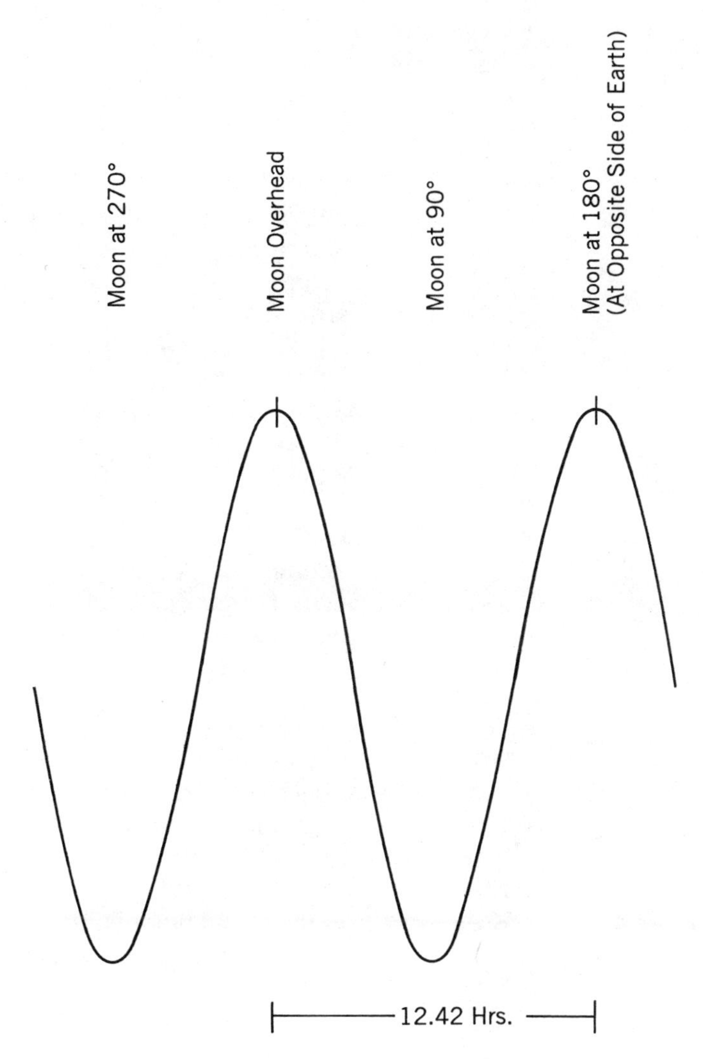

Figure 1.2 Typical semidiurnal tide.

waves comprising a typical tide. These figures show predicted tides for 2-day periods in the Shell Point, Florida, area. For each 2-day period, a prediction was made assuming that only one of the five greatest components (of the tides in this area) was causing the tide. These components included the M_2 constituent, which is the main lunar semidiurnal component with a period of 12.42 hours; the S_2 constituent, which is the main

Table 1.1 Principal tidal constituents.

Symbol	Period (days)	Description
S_a	364.96	Solar annual
S_{sa}	182.70	Solar semiannual
M_m	27.55	Lunar monthly
M_{sf}	14.77	Lunisolar synodic—semifortnightly
M_f	13.66	Lunar fortnightly
$(2Q)_1$	1.167	Second-order elliptical lunar
Q_1	1.120	Larger elliptical lunar
ρ_1	1.113	Larger evectional
O_1	1.076	Principal lunar
M_1	1.035	Smaller elliptical lunar
P_1	1.003	Principal solar
S_1	1.000	Radiational
K_1	0.997	Principal lunar-solar
J_1	0.962	Elliptical lunar
$(OO)_1$	0.929	Second-order lunar
$(2N)_2$	0.538	Second-order elliptical lunar
μ_2	0.536	Variational
N_2	0.527	Larger elliptical lunar
ν_2	0.526	Larger evectional
M_2	0.518	Principal lunar
λ_2	0.509	Smaller evectional
L_2	0.508	Smaller elliptical lunar
T_2	0.501	Larger elliptical solar
S_2	0.500	Principal solar
R_2	0.499	Smaller elliptical solar
K_2	0.499	Declinational lunar-solar
$(2SM)_2$	0.484	Shallow-water semidiurnal
$(2MK)_3$	0.349	Shallow-water third-diurnal
M_3	0.345	Lunar parallax
$(MK)_3$	0.341	Shallow-water third-dirunal
$(MN)_4$	0.261	Shallow-water fourth-diurnal
M_4	0.259	Shallow-water overtide
$(MS)_4$	0.254	Shallow-water fourth-diurnal
S_4	0.250	Shallow-water overtide
M_6	0.173	Shallow-water overtide
S_6	0.167	Shallow-water overtide
M_8	0.091	Shallow-water overtide

solar semidiurnal constituent with a period of 12.00 hours; the N_2 constituent, which is the lunar component due to monthly variation in the moon's distance with a period of 12.66 hours; the O_1 constituent, which is the main lunar diurnal component with a period of 25.88 hours; and the K_1 constituent, which is the solar-lunar component due to declination with a period of 24.07 hours.

At the bottoms of Figures 1.3 and 1.4 are tide predictions using all five of the major constituents for this area. The bottom curves resemble closely the actual tide observed, barring unusual meteorological conditions. The remainder of the 37 constituents typically used for predicting the tides in this country are relatively insignificant in this geographic area for purposes of this illustration. Figures 1.3 and 1.4 also graphically illustrate the previously mentioned spring tide phenomenon occurring when the sun's and moon's constituent cycles are in phase. Note that in Figure 1.3, which is at the time of the new moon and spring tide, the peaks of the M_2 and S_2 cycles occur at roughly the same time. In Figure 1.4, which is at the time of the third quarter moon, the peaks of the M_2 and S_2 cycles are not coincident in time. It may be seen that the composite or actual tide, as reflected by the bottom cycle, is significantly higher in Figure 1.3 than in Figure 1.4 due to the "in-phase" condition.

Occasionally, the spring tides take place at the time of the moon's perigee (when the moon is closest to the earth). When this happens, the highest water of the cycle resulting from the moon's proximity to earth is in phase with (and therefore added to) the highest points of the waves resulting from the moon's and sun's apparent rotation around the earth. This produces abnormally high tides that have been associated with historic coastal flooding (Wood 1976).

The preceding discussion regarding spring and neap tides applies in most areas of the world where *semidiurnal,* or twice daily, tidal oscillations predominate. In some areas, however, only one high water and one low water per tidal day are experienced for most of the month. In such *diurnal tide* areas, tides are more affected by the declination of the moon and sun than by the moon's phases. Declination, in this context, is the apparent angle of an astronomic body above or below the plane

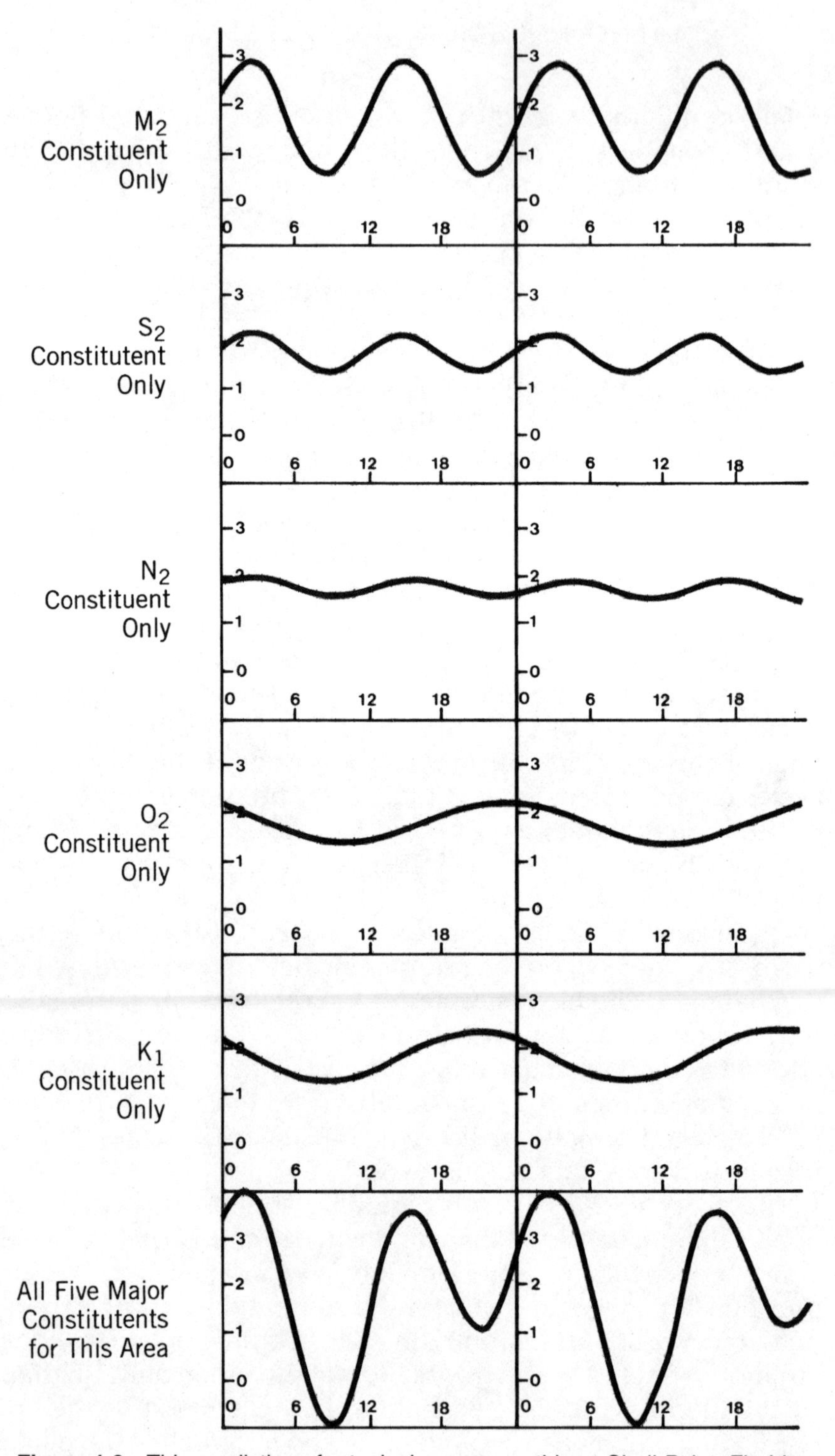

Figure 1.3 Tide predictions for typical new moon tide at Shell Point, Florida.

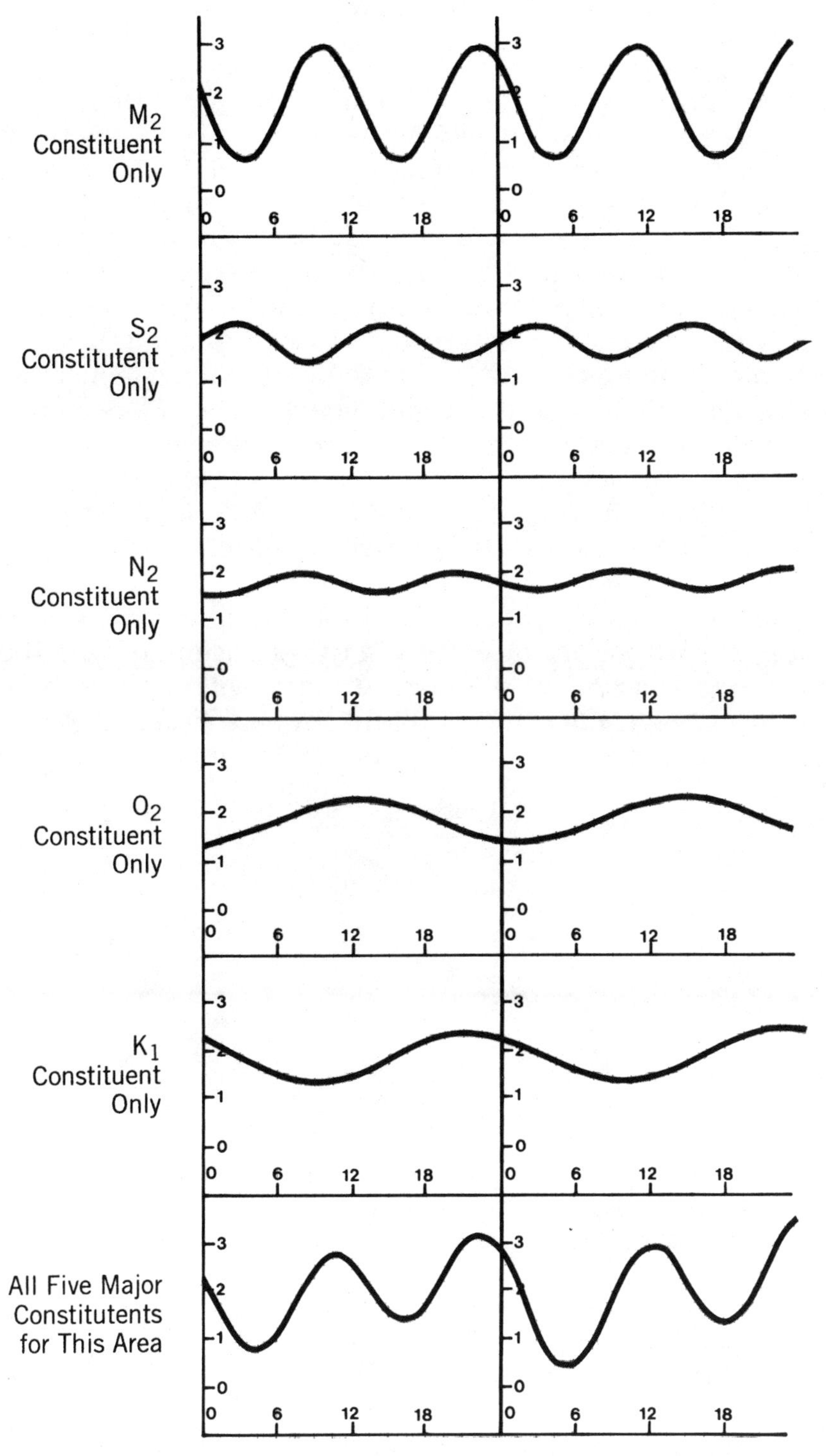

Figure 1.4 Tide predictions for typical third-quarter moon tide at Shell Point, Florida.

of the earth's equator. In such diurnal tide areas, tides with greater range, called *tropic tides,* occur every 2 weeks when the moon is at its maximum northing or southing or maximum declination. When the moon is over the equator, tides of lesser range, called *equatorial tides,* occur. In addition, two high and two low waters generally occur in diurnal areas during such times when the moon is over the equator. The sun's declination has a similar, although lesser, effect on the tides with a maximum near the time of the winter and summer solstices when the sun is at maximum declination and at a minimum during early spring and fall when the sun is over the equator.

The purpose of considering the observed tide as a sum of constituent components is that it allows prediction of tides. If the amplitude and phase lag of any significant constituents are known for a given location, that information and knowledge of certain astronomic factors allow the prediction of both the amplitude and time of the tides at that location for years in advance. The height *(h)* of the tide at any time *(t)* may be represented harmonically by the following formula (Schureman 1958):

$$h = H_0 + \sum_{i=1}^{37} f_i H_i \cos[a_i t + (V_0 + u) - \kappa_i] \tag{1.1}$$

where

H_0	= mean height of water level above prediction datum
H_i	= mean amplitude of constituent i
a_i	= speed of constituent i
f_i	= factor for reducing H_i to prediction year
$V_0 + u$	= equilibrium argument of constituent i at time t
κ_i	= phase angle of constituent i
t	= time reckoned from beginning of prediction year

In this equation, the speed a_i of each constituent is known, as a function of the period; and the factor f and $V_0 + u$ are available from astronomic data tables. The mean sea level, H_0, is site-specific information that may be obtained by averaging periodic tide observations, typically, hourly, over a complete tidal epoch of 19 years (18.6 years, rounded to the nearest whole year to include an integral multiple of the annual constituent associated with the declination of the sun). The amplitude (H_i) and phase angle (κ_i) of each constituent is also site-specific and may be obtained by a process of harmonic analysis of tidal observations. As this process is currently performed, it consists of a least squares analysis of a series of simultaneous equations.

Obviously, the advent of modern computers has made both tidal harmonic analysis and tidal predictions far less laborious. Lord Kelvin, who developed the tidal harmonic constituent analysis theory in 1867, also designed a tide predicting machine. That device was, in essence, a mechanical computer. It physically summed the amplitudes of the harmonic constituents, in accordance with the prediction equation provided before, and traced the resulting curve. Several similar devices were subsequently constructed and were widely used until well into the second half of the twentieth century. The author has a graphic memory of being introduced, as a new employee of the U.S. Coast and Geodetic Survey, to the tide predicting machine being used by that agency. That operation is now easily accomplished on a personal computer with far greater precision.

1.3.2 Tidal Datum Planes

A *tidal datum* is a plane of reference for elevations that is based on average tidal height. Considering the preceding discussion, it is obvious that to be statistically significant, a tidal datum should include all periodic variations in tidal height. Therefore, a tidal datum is usually considered to be the average of all occurrences of a certain tidal extreme for a period of 19 years (18.6 years, rounded to the nearest whole year to include a multiple of the annual cycle associated with the declination of the sun). Such a period is called a *tidal epoch.*

As examples of tidal datum planes, *mean high water* (MHW) is defined as the average height of all the high waters occurring over a period of 19 years. Likewise, *mean low water* (MLW) is defined as the average of all of the low tides over a 19-year tidal epoch. *Mean tide level* (MTL) or *half tide level* is the plane halfway between mean high and mean low water that is used for datum computation purposes. This should not be confused with *mean sea level* (MSL), which is defined as the average level of the sea as measured from hourly heights over a tidal epoch. The relationship between mean sea level and mean tide level varies from location to location, depending on the phase and amplitude relations of the various tidal constituents at each location.

In addition to the preceding planes, there are two other datum planes of significance to water boundaries. *Mean higher high water* (MHHW) is the average of the higher of the high tides occurring each day. *Mean lower low water* (MLLW) is the average of the lower of the low tides occurring each day. Both of these averages are calculated over a tidal epoch. The difference between mean higher high water and mean high water is called *diurnal high water inequality* (DHQ). The difference between mean lower low water and mean low water is called *diurnal low water inequality* (DLQ). These data are shown in Figure 1.5.

From the forgoing, it may be seen that the primary determi-

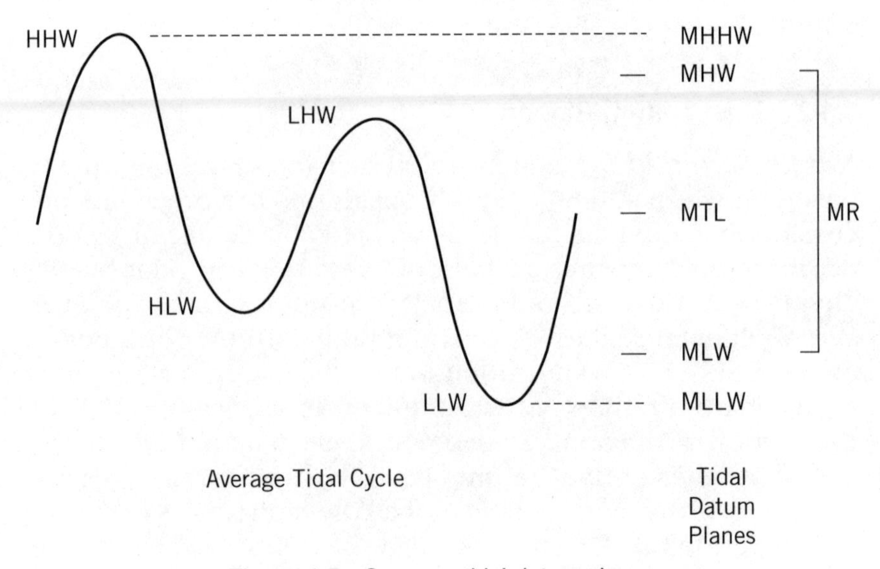

Figure 1.5 Common tidal datum planes.

nation of a tidal datum involves the relatively simple determination of the arithmetic mean, or average, of all the occurrences of a certain tidal extreme over a 19-year tidal epoch. In practice, this is usually accomplished by computing mean values of the various tidal extremes for each calendar month and then annual mean values by averaging the 12 monthly means for each extreme for each calendar year. Finally, the mean values for the tidal epoch used are determined by averaging the annual mean values for the 19 years comprising the epoch. It should be noted that the tidal extremes used in calculation of tidal data are not necessarily the highest and lowest water that occur in a tidal cycle. In most areas, especially on the open coast, wind wave action causes a constant and frequent variation in water level. For consistency, because wind wave action is essentially unpredictable, tidal gauges measure the height of stilled water, which is approximately halfway between the crest and trough of the wind waves. In areas with sizable wind waves, there can be a significant vertical distance between the crest (or trough) of the wind wave and the stilled water level of high and low water.

1.3.3 Local Variation (Cole 1996)

A tidal datum is a local phenomenon because of numerous local topographic forces shaping the tide. In large water bodies, such as the oceans, there is direct response to the tide producing forces in the form of oscillations. The pattern and magnitude of such oscillations are governed by the volume of water available as well as the natural oscillation period of the ocean basin. A natural oscillation period similar to the period of the gravitational force results in reinforcement of the tidal oscillation. The topography of the basin also has an effect on the stilled water level, which in turn affects the average tidal heights in some regions. As an example of this phenomenon, the sea surface has been noted to be depressed as much as 192 feet over ocean trenches and may bulge as much as 16 feet over seamounts (Duxbury 1989).

As the tidal oscillations approach land masses, the shoaling of the continental shelf tends to increase the tidal range. Where openings occur in the coastline, progressive tidal waves travel into such openings, where they are shaped by local topography. Great differences of tidal ranges may result within such openings, as in the Bay of Fundy in southeastern Canada. The average tidal range is 7 feet at the entrance to the Bay and increases to almost 36 feet at the head of the Bay.

As mentioned, once an ocean tidal wave enters an estuary, local topography, rather than astronomic forces, becomes the primary force controlling the wave. One of the leading factors is the location of the head of the bay or other barrier that creates a reflection of the incoming tidal wave. Tidal wave height in a channel at any given time may be considered to be a product of two waves of similar wavelengths and periods but traveling in opposite directions. One of these is the primary wave originating in the open sea and moving up the estuary. The other is a reflected wave originating at a barrier or partial barrier, often at the head of the estuary, moving back down the estuary (Redfield 1950). The resultant wave height at any location is the sum of the two waves.

The net amplitude of the combined waves would be expected to be at a maximum at a point where the direct and reflected tidal waves are "in phase." Because the highest portion of both waves occur at the same time as do the lowest portions of both waves when the waves are in phase, this results in a relatively great observed range. At points when the direct and reflected waves are out of phase, one wave would be at its maximum while the other is at its minimum, and the waves tend to cancel each other. Therefore, at such points, a relatively small observed range results.

The impact of reflected waves on tidal range in estuaries is illustrated by Figures 1.6 and 1.7. These figures provide profiles of the mean tidal range in St. Johns River (Florida) and Chesapeake Bay. The sinusoidal pattern of the tidal range variation is related to the combination of the direct and reflected waves. In St. Johns River, the river is relatively narrow between the ocean and Jacksonville. It then gradually widens and remains a very wide estuary between Jacksonville and Palatka, where it narrows dramatically. The sudden narrowing of the river acts as a partial reflection point that creates a reflected tidal wave. Based on the previous discussion, a minimum point in tidal range would be expected at the point where the direct and reflected waves are out of phase. That point would be one-quarter of a tidal wave downriver from the bar-

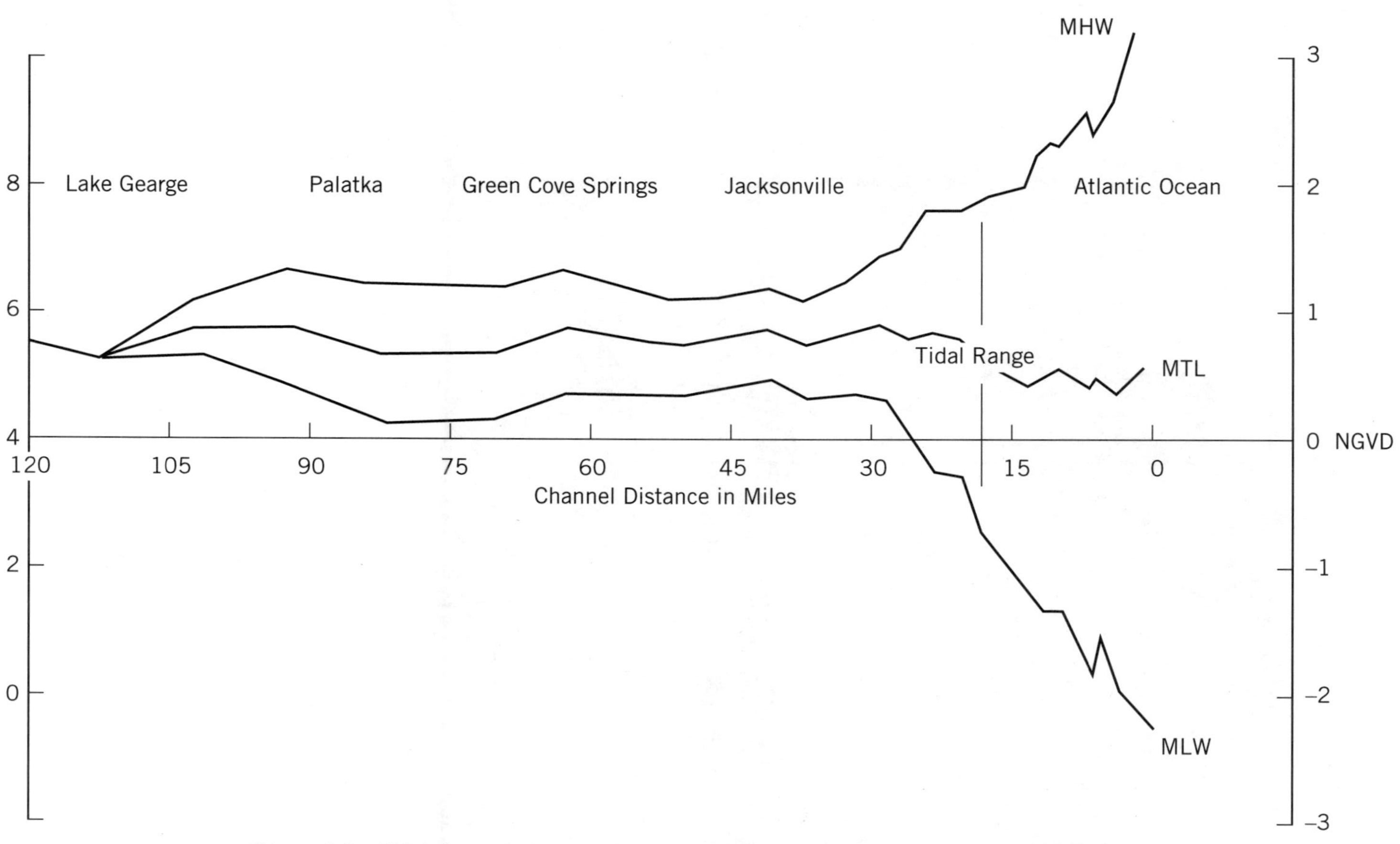

Figure 1.6 Tidal datum heights in feet vs. channel distance in miles, St. Johns River.

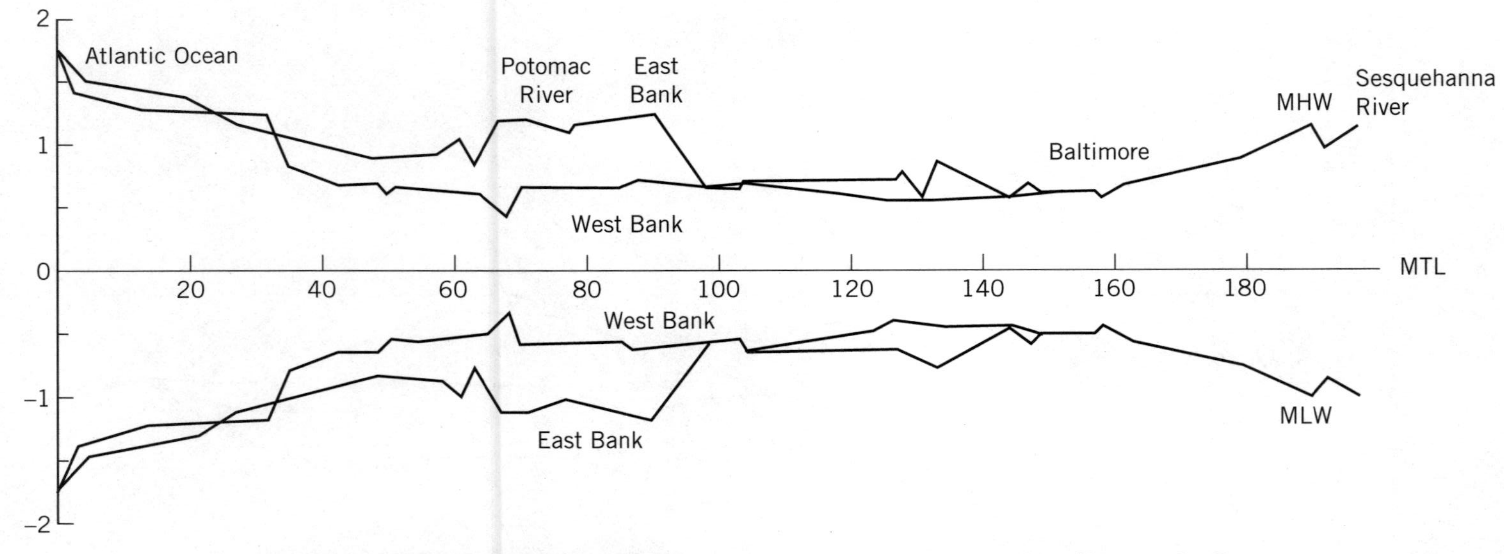

Figure 1.7 Tidal datum heights in feet vs. channel distance in miles, Chesapeake Bay.

rier. At the average rate of travel of the tidal wave in St. Johns River (12 mph), that distance would be approximately 37 miles, which fits the minimum range point, on Figure 1.6, just below Jacksonville.

In Chesapeake Bay, the reflecting barrier is the head of the bay near Harve de Grace. The minimum range points may be noted in Figure 1.7 near mile 125 and 70, which correspond with the lengths of one-quarter and three-quarters of a tidal wave in the Chesapeake. Thus, the inflection points in estuaries may be generally predicted with a knowledge of the travel rate of the tidal wave in the estuary and of the topography of the estuary.

Another shaping factor within estuaries is the Coriolis effect. In wider estuaries, greater tidal ranges are experienced in areas closer to the right-hand side of the estuary when traveling away from the ocean. Such variation is due to the Coriolis effect, which causes a right-hand veer to ocean currents in the Northern Hemisphere (Williams 1962). For the incoming tidal wave, the right-hand veer results in a piling up of water on the right side of the estuary (when traveling away from the ocean) and this creates a slope in the water level across the channel, with the greater high tide experienced on the right side. On the outgoing tide, a similar right-hand veer (to the water mow flowing in the opposite direction) results in an opposite slope and a lower low tide on the right side when traveling away from the ocean (Defant 1958).

The impact of the Coriolis effect is shown in Figures 1.7 and 1.8, where it may be seen that there are differences between the tidal range along the two shores of the Chesapeake Bay and the Bay of Fundy. The larger range is associated with the eastern shore in both estuaries, which is the right-hand side when traveling away from the ocean.

Other factors shaping the tidal range within estuaries include changing width, changing depth, the presence of tributary or distributary water bodies, the distance from the ocean, and changes in the course of the estuary. Based on the principle of conservation of mass, an increase in tidal range is expected in portions of the estuary with decreasing width. Essentially, as the same mass of water in the incoming tidal wave is forced through a smaller channel, greater wave height results.

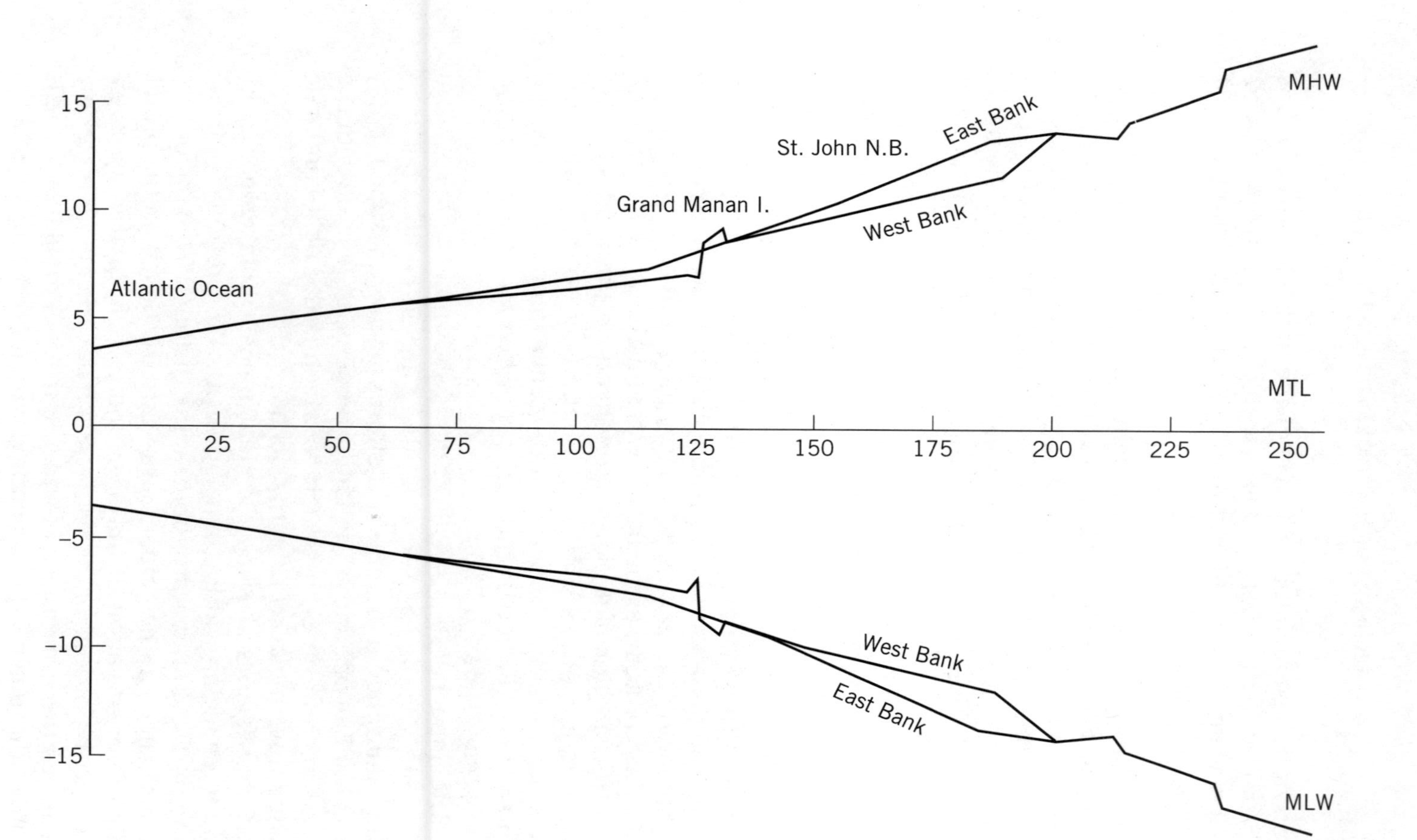

Figure 1.8 Tidal datum heights in feet vs. channel distance in miles, Bay of Fundy.

Conversely, a decrease in range is expected with increasing width. Similarly, an increase in tidal range is expected in portions of the estuary with decreasing depth. Conversely, a decrease in range is expected with increasing depth. Also based on the principle of conservation of mass, a decrease in tidal range is expected in portions of the estuary near large tributaries (or more specifically distributaries) that would receive a portion of the mass of the incoming tidal wave. Because a portion of the water mass is diverted into any distributaries, this would be expected to result in less height for the tidal wave in the main channel.

Increasing distance from the ocean has a decreasing effect on tidal range due to friction. However, many estuaries tend to narrow and become more shallow in their upper reaches, so the tidal range often tends to increase with distance. Considering the basic principles of fluid mechanics regarding frictional drag, it is expected that due to frictional losses associated with curved channels, portions of estuaries with large percentages of curvature are expected to have reductions in tidal range.

It may be seen from the preceding that there are a number of different topographic factors that shape the incoming tidal wave, resulting in significant differences in elevation of a tidal datum from point to point in even the same general vicinity (Cole 1977). Therefore, a tidal datum should be determined in the immediate area of its intended use.

1.3.4 Sea-Level Changes

As mentioned in the previous section, a tidal datum is defined as an average over a 19-year period known as a tidal epoch. Traditionally, for all datum values published by the National Ocean Service, all data are referred to a specific epoch called the *National Tidal Datum Epoch.* A specific 19-year period is used because apparent nonperiodic variation in mean sea level is noted from one 19-year period to another. It is not known if these trends are truly nonperiodic or a part of some long-term oscillation. These apparently nonperiodic changes are possibly due to "glacial-eustacy, thermal volumetric changes, vertical land movements, and both climatological and oceanographic trends" (Hicks et al. 1983).

During the last 100 years, sea-level monitoring has indicated a worldwide trend of continual rise in sea level. To correct for this rise in the United States, a new epoch has been historically adopted every two or three decades when significant change has occurred. At such times, adjustments are made to all datum elevations. In effect, a quantum jump occurs in the elevations of all tidal datum planes for stations published by the National Ocean Service at those times. An example of this occurred in 1981 when a change in epoch was adopted. Previously, the epoch of 1941–1957 was used. The National Tidal Datum Epoch adopted in 1981 is 1960–1978.

In recent years, sea level has been rising at an average rate of 0.0066 foot/year in the United States (Hicks and Hickman 1988). Some sections of the coast have a much higher rate. An example of this is the coast of Texas. Figure 1.9 shows a plot of sea level rise for Freeport, Texas, which has an average rise of 0.046 foot/year (Lyles, Hickman, and Debaugh 1988).

Obviously, in areas such as the Texas coast with above-average rates of sea-level rise and even in areas with average rates of sea-level rise, a significant difference can exist between an elevation computed in the National Tidal Datum Epoch and a datum computed in the most recent 19 years. Therefore, it may be more desirable to recompute data in a more current epoch than use published data computed in the National Tidal Datum Epoch.

1.3.5 Datum Computations

As previously discussed, the primary determination of a tidal datum involves the relatively simple calculation of the arithmetic mean, or average, of all the occurrences of a tidal extreme over a 19-year tidal epoch. Most tidal datum elevations, however, are determined from observations of less than 19 years. Methods have been developed for correcting such short-term observations at the desired point and at a control station at which 19-year mean values are known. The average of the observed tidal extremes then may be reduced to a value equivalent to a 19-year mean by a correlation process using a ratio of tide ranges observed at the two stations. Methodology for this is as follows:

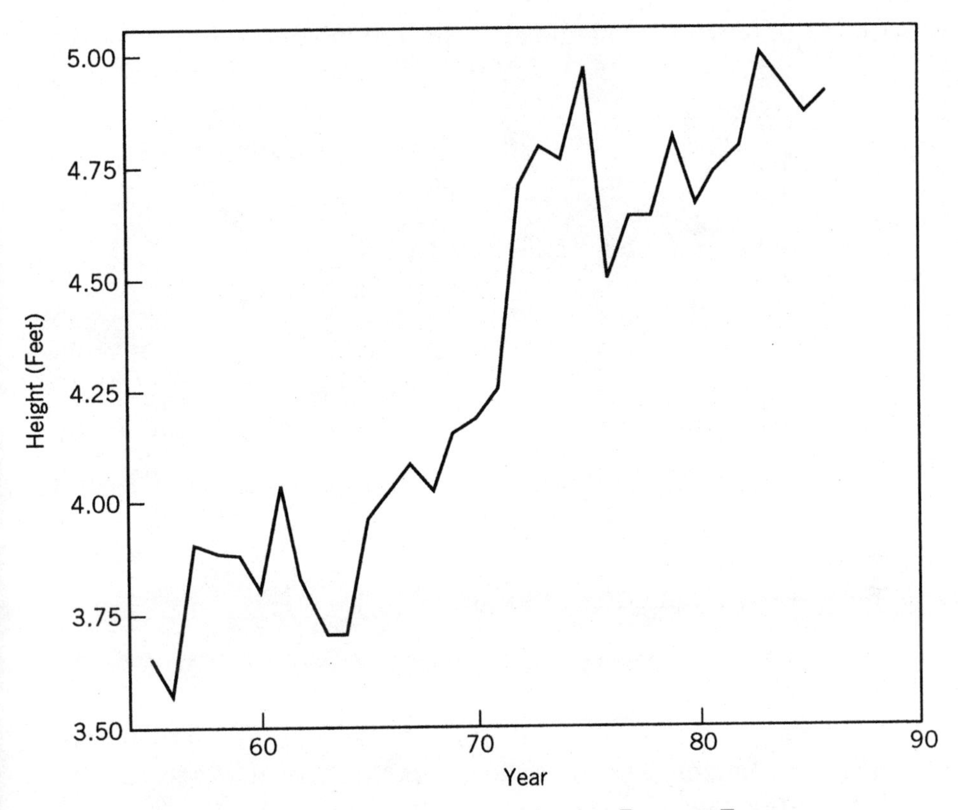

Figure 1.9 Plot of annual mean sea level at Freeport, Texas.

Standard Method. The Standard Method (Marmer 1951) for accomplishing this process, also known as the range ratio method, is based on the following equations,

where

MHW = 19-year mean high water
MTL = 19-year mean tide level
MLW = 19-year mean low water
MR = 19-year mean range
TL = mean tide level for the observation period
R = mean range for the observation period
s = subscript used to denote the subordinate station
c = subscript used to denote the control station

The Equation (1.2) calculates the equivalent 19-year mean range at the subordinate station:

$$\frac{MR_S}{R_S} = \frac{MR_C}{R_C} \tag{1.2}$$

This equation may be restated as follows:

$$MR_S = \frac{(MR_C)(R_S)}{R_C} \tag{1.2a}$$

The Equation (1.3) calculates the subordinate 19-year mean tide level:

$$TL_C - MTL_C = TL_S - MTL_S \tag{1.3}$$

This may be restated as follows:

$$MTL_S = TL_S + MTL_C - TL_C \tag{1.3a}$$

Equations (1.4) and (1.5) calculate the 19-year mean high and mean low water by applying half of the mean range to the mean tide level:

$$MHW_S = MTL_S + \frac{MR_S}{2} \tag{1.4}$$

$$MLW_S = MTL_S - \frac{MR_S}{2} \tag{1.5}$$

For determining mean higher high water and mean lower low water, the Standard Method uses the following equations, where

MHHW = 19-year mean higher high water
MLLW = 19-year mean lower low water
MDHQ = 19-year mean diurnal high water inequality
MDLQ = 19-year mean diurnal low water inequality
DHQ = Diurnal high water inequality for the observation period
DLQ = Diurnal low water inequality for the observation period

Equations (1.6) and (1.7) calculate the equivalent 19-year mean diurnal high and diurnal low water inequalities at the subordinate station:

$$MDHQ_s = \frac{(DHQ_s)(MDHQ_c)}{DHQ_c} \tag{1.6}$$

$$MDLQ_s = \frac{(DLQ_s)(MDLQ_c)}{DLQ_c} \tag{1.7}$$

Equations (1.8) and (1.9) calculate the equivalent 19-year mean higher high and lower low water at the subordinate station:

$$MHHW_s = MHW_s + MDHQ_s \tag{1.8}$$

$$MLLW_s = MLW_s - MDLQ_s \tag{1.9}$$

The National Ocean Service recommends that a procedural departure to the Standard Method be followed if the tide at the control station is predominately diurnal and there is a significant difference in the total number of highs and lows between the control and subordinate stations (National Ocean Survey 1980). That method is as follows: "The mean of the diurnal high water inequalities would be computed *without adjustment* at the subordinate (1-year minimum) station. This value would then be subtracted from the adjusted (to the control station) Mean higher High Water to give the datum of Mean High Water at the subordinate station. Likewise, the unadjusted diurnal low water inequality would be added to the 19-year adjusted value of Mean Lower Low Water to give Mean Low Water at the subordinate station."

It is noted that the use of regression analysis has been suggested as a substitute for the standard equations provided before (Cole, Speed, and Fugate 1989). Such methods perform a more statistically valid correlation as well as providing a means of evaluating the precision of the results. Preliminary tests indicate that it is a promising method for accomplishing secondary determinations.

The Standard Method requires observation of both the high and low tidal extremes at the control and subordinate stations during the period of observation. In recent years, a demand has developed for determination of mean high water elevations in areas where only the upper portion of most tidal cycles are observable. These are areas such as marshes, mud flats, and tidal creeks with wide, flat, intertidal zones. Two alternate methods are in current use for such determinations, the Amplitude Ratio Method and the Height Difference Method.

Amplitude Ratio Method. The Amplitude Ratio Method (Cole 1981) was derived to mathematically duplicate the results of the Standard Method when only the top portion of the tidal cycle is available at the unknown or subordinate station. This method recognizes that for two sine waves of similar wavelength, the differences in height between the peaks of the cycles and the points at which the curves "fit" time periods of equal length are proportional to the ratio of the two amplitudes. It is noted that this method requires a computation on each observed cycle rather than using mean data for the entire observational period. An average of the results from the individual computations for the period of observation is usually used for the final value.

This method (see Figure 1.10) may be defined by the following equations:

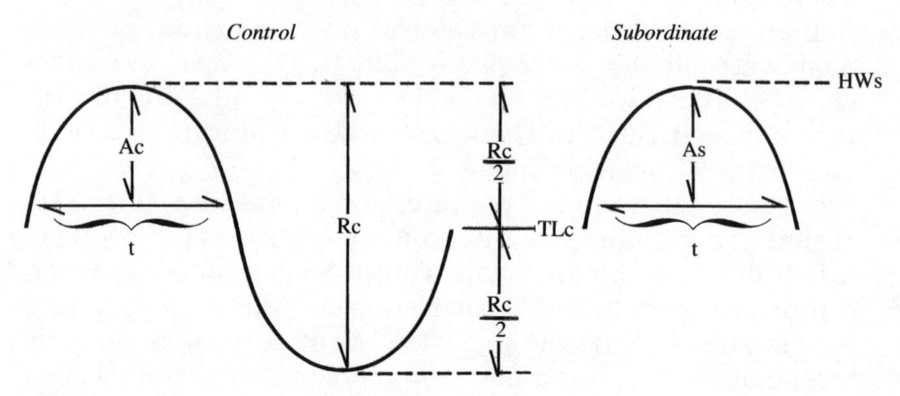

Figure 1.10 Computation procedures (Cole 1981) amplitude ratio method.

Equation (1.10) computes what the observed range at the subordinate station would have been if the entire cycle could have been observed:

$$R_S = \frac{R_C A_S}{A_C} \tag{1.10}$$

where

A = observed interval between peak high water and the elevation on the tide curve at the selected time interval

Equation (1.11) computes the equivalent 19-year mean range at the subordinate station:

$$MR_S = \frac{(MR_C)(A_S)}{A_C} \tag{1.11}$$

Mean tide level, mean high and mean low water then may be calculated by Equations (1.3) to (1.5) of the Standard Method.

If it is desired to compute mean higher high water by use of observations of partial tidal curves, elevations should be determined for the peak high waters for two consecutive tidal cycles occurring in a tidal day. The higher of the high waters represents higher high water and the other peak the lower high water. The difference in elevation between the two represents the diurnal high water inequality (DHQ) for the day. By using Equations (1.6) and (1.8) and these observed values, the equivalent 19-year mean DHQ and mean higher high water may be determined.

Height Difference Method. The Height Difference Method (Swanson 1974) is the second alternate method (see Figure 1.11). This method also gives results comparable to the Standard Method in areas where there is little or no difference in range between the control and subordinate station or where the period of observation covers a substantial period of time. This method assumes that the elevation difference between mean high water and peak high water is the same at both the control and subordinate stations for a given tidal cycle. For

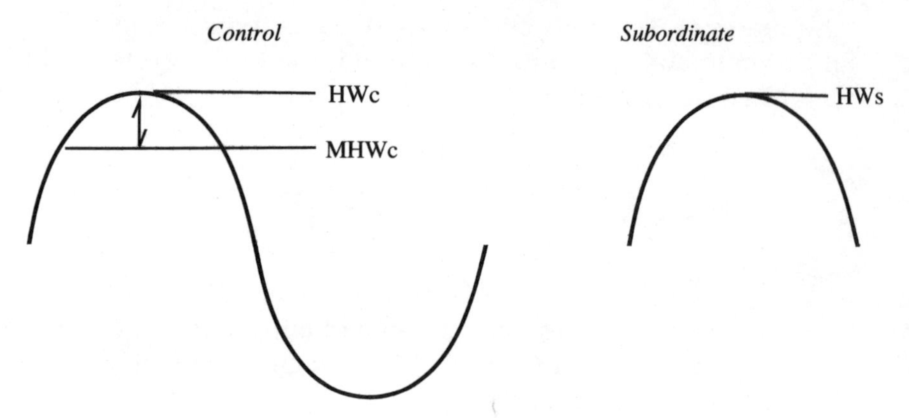

Figure 1.11 Computation procedures height difference method.

observations of more than one cycle, it assumes that the elevation difference between 19-year mean high water and the mean high water for the period of observation is the same at both stations. In this method, the height difference between observed high water and the known 19-year mean high water at the control station is applied to the observed high water at the subordinate station to estimate the 19-year mean high water at that location by the following equation:

$$MHW_s = HW_s + MHW_c - HW_c \tag{1.12}$$

where HW is the mean high water for the observation period.

When there is a substantial difference in range between the control and subordinate stations, this method can result in an incorrect estimate of mean high water at the subordinate station. However, if the range ratio between the control and subordinate stations has previously been determined by the Standard Method or Amplitude Ratio Method, the following equation may be used to compute an additive correction to the value determined by the Height Difference Method (Zetler 1981):

$$\text{Correction} = \frac{(1 - AR)(R_c - MR_c)}{2} \tag{1.13}$$

where

$$AR = \frac{R_s}{R_c}$$

Mean higher high water may be calculated by a variation of this method by substituting $MHHW_c$ for MHW_c in Equation (1.12) and by substituting observed and mean diurnal range for observed and mean range. (Diurnal range is defined as the vertical interval between mean higher high water and mean lower low water.)

Constituent Analysis Method. The three previously discussed methods for determining tidal datum planes involve observation of high and low extremes of the daily, or twice daily, tidal cycles and the averaging of the heights of these tidal extremes. This traditional approach is demanding on data acquisition in that it requires a sampling of the water level at the time of the tidal extremes, which implies continuous sampling to ensure the capturing of that instant. If the water-level gauge malfunctions during the needed times, considerable data may be lost. Such an approach is wasteful of data because, although continuous data must be collected, only the extreme values are utilized. Furthermore, only the extreme values are used, so observations must be taken for a significantly longer period to obtain statistical accuracy than would be the case if all data could be utilized. Prior to the ready availability of computers, this traditional approach was desirable due to the inherent simplicity of the calculations. With the proliferation of computers and modern analysis techniques, however, alternative approaches are available.

Tidal datum elevations also may be calculated by using the amplitudes of the harmonic constituents making up the observed tide. Such an approach arguably may result in statistically more valid determinations with comparable data; and such an approach is considerably less demanding of data acquisition because continuous data are not necessary.

Once constituent amplitudes are determined through the previously discussed harmonic analysis process, datum elevations may be readily calculated. Although theoretical approaches may be used for this determination, a more direct approach is to perform a prediction of tides over a complete tidal epoch of 19 years and average the predicted twice daily tidal extremes. Typically, tidal predictions require knowledge of both the amplitude and phase of the harmonic constituents. For the purpose of datum determination, however, there is no concern for the temporal accuracy of the predictions. Therefore, a simplified prediction may be used in which the initial phase relationship between the constituents is not required. That simplified prediction may be described mathematically as follows:

$$h = H_0 + \sum_{i=1}^{37} f_i H_i \cos a_i t \tag{1.14}$$

H_0 = mean height of water level above the prediction datum
H_i = mean amplitude of constituent i
a_i = speed of constituent i
f_i = factor for reducing H_i to the prediction year
t = time reckoned from beginning of the prediction year

With tidal observations of less than a complete 19-year tidal epoch, the harmonic analysis process should include corrections to the constituent amplitudes for the 19-year nodal cycle. For observational periods of less than a year, approximation of a primary determination may be made by using amplitudes for the shorter-period constituents derived by harmonic analysis of observations at the subject site and amplitudes for the longer-period constituents that have been derived at suitable nearby control stations.

Alternative Methodology. All of the preceding datum computational methods, including the constituent analysis method, are based on the Anglo/American common law definition of a tidal datum, which uses the average of the twice daily tidal cycles. Use of such methods ignores all variations with periods greater than one-half day. Such a datum may be appropriate in the British Isles, where it originated, because the only significant tidal signal is the twice daily cycle. However, the use of tidal datum planes based on averages of the twice daily tidal cycles may be inappropriate in areas where a significant portion of the tidal signal involves constituents with periods in excess of one-half day. For example, in areas with significant daily inequality in tidal cycles, the physical shoreline may be shaped more by the greater of the two daily tides than by a fictitious average tide. In such areas, the higher high water therefore may be a more appropriate datum for boundary purposes from the scientific perspective. Similarly, in areas with pronounced seasonal tidal constituents, the physical shoreline may be shaped more by the semiannual or annual tidal cycles, which suggests a semiannual or annual high water as the most appropriate datum for boundary purposes. Although such an approach would obviously result in a boundary more in accord with physical shoreline features, it should be noted that these observations are solely from a scientific perspective. Such an approach has not been subjected to a judicial test to the knowledge of the author.

In areas with significant long-term tidal constituents, a datum that includes consideration of those constituents, may be determined by a slight modification of the previously described constituent analysis approach. For example, in areas with pronounced seasonal variation in the tides, a *mean annual high water line* might be most appropriate, especially in areas where civil law controls. To derive a datum for such a line, predictions could be made with Equation (1.14) for any 19-year period and an average determined of the predicted annual high waters.

1.3.6 Tide Gauging Techniques

Most tide observations of duration longer than a few hours are made on recording tide gauges that either continuously or at fixed intervals record the water level. The simplest of these are operated by a float that moves up and down with the rise and fall of the water level in a stilling well. The vertical movement of the float moves a pencil back and forth across a strip of paper driven by a clock mechanism, resulting in a graph of the rising and falling tide. A refined version of this gauge is widely used that punches the water-level reading in binary

Figure 1.12 Analog-to-digital gauge.

code onto paper tape ten times an hour. This punched tape output allows for easy processing of the resulting data by computer. Figure 1.12 shows such a gauge.

Other types of gauges use pressure sensors or measure the water height by timing sound waves or laser beams bouncing off the water surface. These have a variety of output devices, usually with magnetic tapes, disks, or computer chips.

Whenever a float mechanism acts as the sensor for the gauge, a stilling well should be used to dampen short-period waves, such as wind waves, and allow only the longer-period tidal waves to move the float. A stilling well is a vertical tube that extends below the lowest possible tide. Water enters the well through an intake near the bottom of the well and rises to the average level of the water surface outside the well. The size of the intake controls the amount of damping that takes place. The opening must be large enough to allow sufficient flow of water for tide measurements while still damping the rapid fluctuation of water level caused by heavy seas. For a 12-inch-diameter well, the National Ocean Service recommends an intake diameter size of 1.5 inches for protected areas and a 3/4- to 1-inch diameter well for areas exposed to open seas (Coast and Geodetic Survey 1965). For smaller-sized wells, the opening should be proportionally sized. Another rule of thumb often used for intake sizing is that the intake area should be 1/50th of the areas of a cross-section of the well.

When recording gauges are used for tide datum determination, an additional complication is introduced in relating the results to a bench mark or ground elevation. This is because most recording gauges cannot be directly leveled to as can a staff gauge. Therefore, a tide staff, usually consisting of a graduated, vitrified, glass scale, is installed near the gauge. Comparisons are made between the gauge and the staff to determine the constant difference in reading between the two. The staff may be referenced by leveling to several permanent bench marks in the area. These bench marks will preserve the elevation of the tidal datum that is determined after the gauge and staff are removed. Figure 1.13 provides a diagram of a typical short-term tide station installation with a float-operated gauge. When tide gauges are installed for permanent operation, a more elaborate installation, protected by an especially built shelter, is often used. Figure 1.14 illustrates the permanent tide station operated by the National Ocean Service at Port Mansfield, Texas.

The National Ocean Service has developed standards for their tide data acquisition (Bodnar 1977). These include the requirement for ten bench marks at permanent tide stations and five at secondary tide stations. Three of these are required to be in bedrock or on stainless steel rods driven to bedrock or refusal. The standards also require that the Second Order, Class I specifications, as defined by the Federal Geodetic Control Commission, be used for leveling at either primary or secondary stations and also require the use of invar-level rods. As may be seen, these standards are quite stringent because they are designed for a large data acquisition network involving numerous personnel and many different locations and circumstances. Depending on the circumstances, these standards may or may not be applicable to smaller tidal studies.

For very short-term observation, a simple graduated staff, which may be read at 5- or 10-minute intervals, is often used.

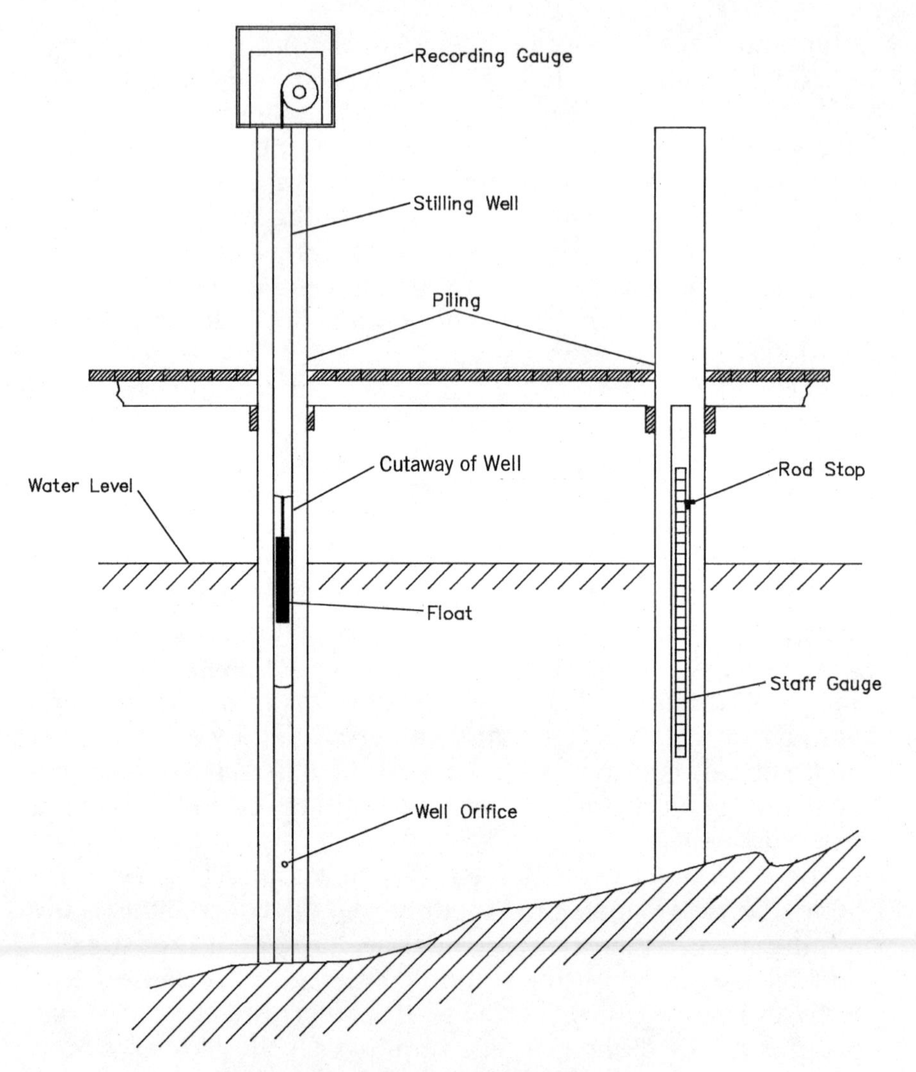

Figure 1.13 Typical short-term tide station.

This offers minimum expenditure for equipment. This also offers a minimum of effort and time needed for installation because the staff may be either driven into the water bottom or secured to a piling. In some areas where it may be desirable to determine a tidal datum (such as intertidal marshlands), such a manually read staff may be the most practical gauge.

Figure 1.14 Typical permanent tide station. (Photo courtesy of the National Ocean Service, NOAA.)

Float-type gauges tend to experience problems in such areas that are dry for a portion of the tidal cycle.

A recent trend in tidal measurement has been toward gauges equipped with telemetry devices to allow near-real-time monitoring of tides in remote locations. The National Ocean Service is in the process of equipping many of its permanent tide stations with telemetry devices that relay the data to their headquarters via satellite. In another example of this trend, Texas, in a program conducted by Corpus Christi State University, has developed a relatively inexpensive adaption to gauges that uses radio telemetry (Cole and Dearsing 1989). Such devices can provide an economical approach to data acquisition, especially in remote areas.

When tide stations are established in organized programs along a coastline, a hierarchy of stations is usually established. This scheme uses long-term (minimum of 19 years of observation) stations as primary control stations. These are spaced typically at 100- or 200-mile intervals. Within the primary control framework, a series of secondary stations are usually operated for at least 1 year. These are spaced with at least one in each bay or in another area with unique tidal characteristics. Tertiary stations, with 30 to 90 days of observations, are used to further densify the coastline between the secondary stations. In areas with such a hierarchy of tide stations, datums between tertiary stations may be generally established, when needed, by only a few cycles of simultaneous observations.

The selection of optimum sites for a network of tide datum stations is a subject that deserves careful attention. When considering the previous section on tidal variation, it may be seen that tidal variation has some degree of predictability. The use of such predictability may allow the selection of gauge sites that minimize the number of gauges required and optimize the resulting definition of the water body.

As an example, the concept of reflected tidal waves appears to be the cause of the sinusoidal tidal range pattern often observed in long, linear estuaries. Based on this concept, the location of the maximums and minimums of such sinusoidal patterns are somewhat predictable. The maximums should be located near the zero, one-half, and full tidal wavelength distances from a reflecting barrier in the channel, and the minimums should be located near the one-quarter and three-quarter tidal wavelength distances from the barrier. This allows for the selection of gauge sites at the predicted location of such inflection points that allow for optimum definition of the tidal datum variation within the estuary. The estimated distance from the reflecting barrier to such inflection points may be calculated with the following equation:

$$\text{Distance (in miles)} = FPS \qquad (1.15)$$

where

F = fraction of tidal wave (1/4 and 3/4 for predicted low points and 1/2 and 1 for high points)
P = tidal wave period (12.42 hours for semidiurnal or mixed and 24.84 hours for diurnal tides)
S = speed (mph) of tidal wave in an estuary

Ideally, gauges should be located near each inflection point as well as at several intermediate points to define the variation in an estuary. In addition, definition may be needed near places where there are features that affect friction (narrows, sudden widening, and so on) and at entrances of tributaries or distributaries. In wide estuaries, lateral slope, caused by the previously discussed Coriolis effect, also must be taken into account in the selection of gauge sites.

As may be seen from Equation (1.15), it is necessary to have knowledge of the general configuration of the estuary and the average travel speed for a tidal wave in the estuary. This may be readily determined by the predicted time lag between the arrival of a tidal extreme at the mouth of the estuary and arrival at a reflecting barrier typically located at the head of the estuary. This information is available for many areas in tide tables published annually by the National Ocean Service.

1.3.7 Interpolation Between Tide Gauging Stations

In many cases where a tidal datum is required in an area with an existing well-planned network of tidal datum points, a precise datum may be established by interpolation, as opposed to supplemental tidal observations at desired locations. On long stretches of regular and unbroken open ocean coastlines, tidal range variation generally is relatively linear. In such areas, linear interpolation is generally an acceptable method of determining a local tidal datum elevation. This may be accomplished by determining the elevation, in relation to a common datum, of the nearest existing tide station on either side of the desired point. Such an approach is not recommended where there are intervening inlets, where either existing tide station is near an inlet, where the coastline is irregular, or where the distance between the two stations is excessive.

Linear interpolation is also generally not recommended within estuaries. Based on Figures 1.6 through 1.8, it may be seen that variation typically is not linear in estuaries. The shape of the variation pattern is somewhat predictable, however, based on the parameters discussed in the previous section on local variation. Therefore, interpolation within estuaries may be a valid option but may require the use of a more complex line form for the interpolation.

In a recent study (Cole 1996), it was found that the variation pattern within estuaries could be precisely described by a polynomial equation fit to the tidal data by a regression analysis process. The resulting polynomial equation then could be used for precise interpolation between existing tide stations. The order of the polynomial equation that best fits each estuary was found to be predictable based on the length of the estuary and the travel speed of the tidal wave within the estuary.

The reason for this predictability is the impact of reflected tidal waves on the incoming tidal wave. As discussed in a previous section, this is one of the major determinants of the pattern of variation. Figure 1.15 shows this impact in estuaries with a theoretical tidal range variation in a uniform channel when plotted against channel length from the mouth of the estuary to its head or other barrier as measured by the portion of a tidal wave that would fit in the estuary. Thus, this measurement of channel length is a function of the speed at which the tidal wave travels the length of the estuary (which is a function of depth and other frictional forces) as well as the physical length itself. This measurement of channel length is typically called *phase.*

Phase may be calculated by first determining the time it takes an average tidal wave to travel the length of the estuary from its mouth to its head or other barrier *(time of transit).* This information is available for many areas in tide tables published annually by the National Ocean Service. The phase then may be calculated by the following equation:

$$\text{Phase} = \frac{(\text{time of transit})(360°)}{\text{tidal wave period}} \qquad (1.16)$$

For this calculation, the tidal wave period is 12.42 hours for semidiurnal tides and 24.84 hours for diurnal tides. Once phase is determined, use of Figure 1.15 predicts the general shape of the variation pattern, although other factors may cause distortions in the general shape. As an example, tide waves in the Bay of Fundy, which has a relatively high tidal

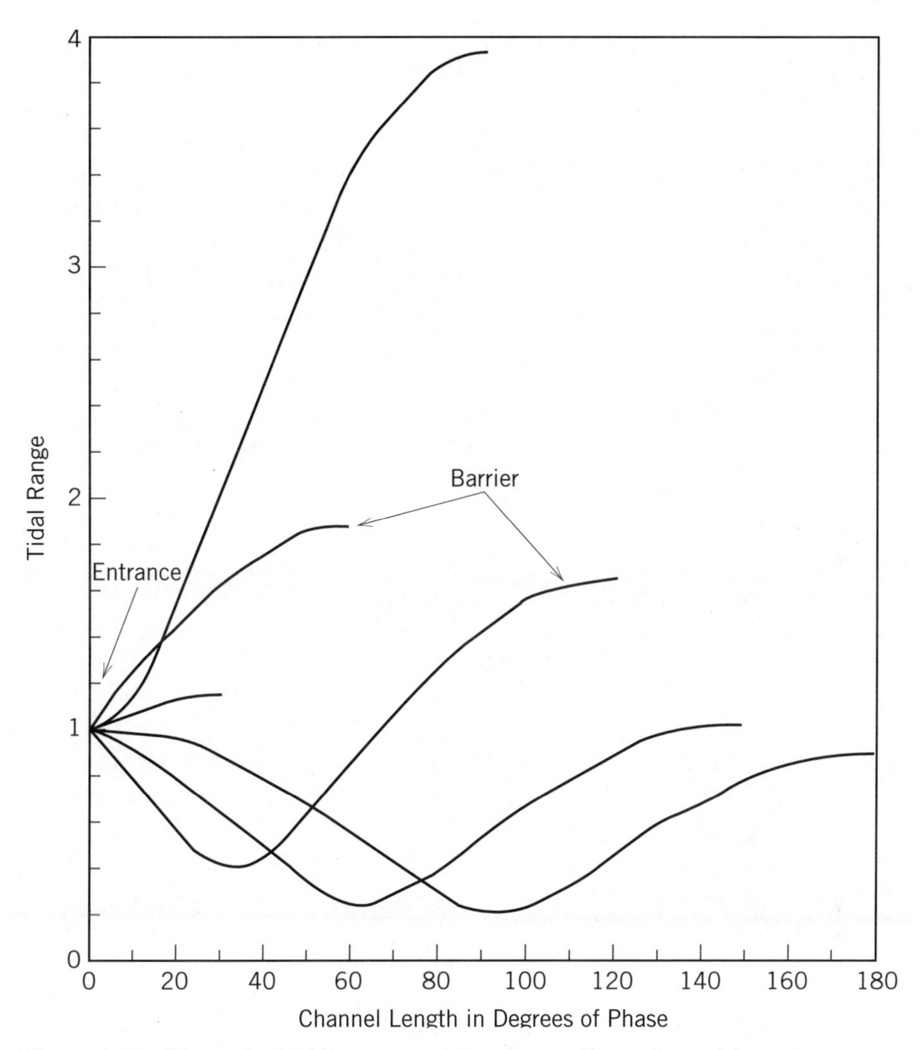

Figure 1.15 Theoretical tidal range variation in a uniform channel based on a range of 1 at the entrance and a uniform damping coefficient of 1.0 (after Redfield 1950).

wave travel speed (approximately 90 knots), has a phase value of 69 degrees with a physical length of over 200 miles. From Figure 1.15, it is seen that the predicted shape is a steadily and rapidly rising curve. From the actual curve for the Bay of Fundy in Figure 1.8, it is seen that it is similar to the predicted shape.

As a second example, tide waves in the St. Johns River travel at a much lower speed (approximately 12 knots), resulting in a phase value of 220 degrees with a physical length of approximately 90 miles to the end of the tidal effect. Based on Figure 1.15, a much more complex curve is predicted, first decreasing and then increasing. From Figure 1.6, the observed tide curve is similar to the predicted shape.

These variation patterns may be considered as polynomial equations whose orders vary with the physical length of the estuary and the speed with which the tidal wave travels through the estuary. An estuary that holds no more than one-quarter of a tidal wave (90-degree phase value) is expected to have a first- or second-order equation, and one that holds a greater portion of a wavelength would be expected to have a third-order equation.

1.3.8 Techniques for Locating Tidal Datum Lines

The distinction between the datum or elevation of a tidal boundary and the tidal datum line is an important concept in water boundaries. A tidal datum remains relatively constant, for practical purposes, at a given location over the years. On the other hand, a tidal datum line, which is the intersection of the datum with the rising land, may vary considerably as the land erodes or accretes under the same elevation of water. Therefore, the line may be ambulatory and its location should be related to a specific point in time.

Once a local tidal datum has been established, there are basically three methods that may be used for locating the corresponding tidal datum line in the area around the datum determination. These are the staking method, the topographic method, and tide-coordinated aerial photography.

Often the most practical of these in areas without significant wave action is the staking method. This method allows the water itself to define the line. As an illustration of this method, assume that the correct reading on a staff for a tidal datum has been previously determined. On a tide that is predicted to reach or exceed that value, the staff is observed. When the water level reaches the predetermined staff reading for the desired tidal datum, a signal is given. At the signal, personnel in the area around the staff place a series of stakes at frequent intervals along the incoming edge of the water. These stakes, defining the tidal datum line, then may be mapped by any of various horizontal surveying procedures. The obvious advantage of this method is that the surveyor sees the line on the ground and can identify all inflection points.

The topographic method consists of assuming that the local datum line is a topographic contour in the immediate area around the datum determination. This contour line is then located by leveling and mapped by horizontal surveying procedures. Caution is necessary with this method because only points on the line are being located as contrasted with a continuous line as in the staking method. Therefore, significant breaks and inflections in the line may be overlooked unless care is taken.

Tide-coordinated aerial photography may be used for mapping a tidal datum line when the water/land interface is not obscured by dense vegetation. This method involves the use of aerial photography coordinated by an observer on the ground watching a tide staff. At the precise time that the tide reaches the predetermined staff reading for the tidal datum, the observer signals the aircraft and the water/land interface is photographed with black-and-white infrared film. The infrared photography graphically depicts the interface between the water and upland if the land is relatively dry.

Another photogrammetric process may be used for mapping the approximate location of a tidal datum line in some areas. This method uses aerial photography for interpolation between known points on the line (Cole 1982). Usually, this is accomplished by placing aerial targets on known ground truth points. The line between these points is then interpolated using the tones and textures seen in the photography. Infrared color, natural color, as well as black-and-white photography have all been used with success in various areas. In many heavily vegetated marshlands, this process may be the most practical means of mapping the line.

As previously discussed, a tidal datum line may be ambulatory. Accordingly, littoral owners may gain or lose land with accretion, relition, and erosion. Exceptions to this general rule are discussed in Chapter 4. There is another possible exception to the general rule that may affect the timing of tidal boundary surveys. This issue deals with beaches that have a predictable and cyclic seasonal shift in the location of a tidal datum line.

One Florida case *(Trustees of the Internal Improvement Trust Fund v. Ocean Hotels, Inc.)* addressed this issue. That case involved the problem of determining a boundary on a beach, "which through the natural processes of erosion and accretion, undergoes a predictable, seasonal loss and replenishment of approximately 90 feet of beach sand." In this case, the court found the fluctuating boundary concept unacceptable as a property law standard. After reviewing three possible options (the landward [winter] mean high water line, the seaward [summer] mean high water line, and the average of the two), the court found the winter line to be the boundary. The other two options were dismissed as violating the public trust lands.

Although the preceding represents one case in one geographic area, it does represent an area of concern that should be considered in surveying these boundaries. Another area of concern should be the definition of the tidal datum line being used. Possibly the most commonly accepted definition of a tidal datum line, such as the mean high water line, is "the intersection of the land with the water surface at the elevation of mean high water" (National Ocean Survey 1975). As may be seen, this definition assumes use of the stilled water level at mean high water. The previously described methods for location of tidal datum lines, with the possible exception of the staking method, are based on that definition. It should be noted that in some locations, use of this definition may not result in a line meeting the intent of early common law. For example, early English common law described sovereign lands as including "the foreshore which is overflowed by ordinary . . . tides" (Hale 1666).

This quote assumes the use of the average *reach* of the water at mean high tide, which may be considerably different in elevation from the elevation of mean high water in some areas. For example, on open ocean coastlines, the average reach of the water at mean high water exceeds the mean high water line (as defined before by the National Ocean Service) by a vertical height that includes one-half the average wind wave height as well as the vertical component of the average run-up (also called setup) of the waves. In other areas, especially in shallow lagoons, wind setup may also cause the average reach of the water to be considerably above the mean high water line (as defined before by the National Ocean Service).

The average wave height may be determined by direct measurement or by use of tabulated values available for many areas. Also, average wave run-up may be calculated by a standard equation used for beach protection design (U.S. Naval Facilities Engineering Command 1982). Further, average wave setup due to winds may be calculated (U.S. Army Corps of Engineers 1977). However, the practicality and legal acceptance of application of such corrections would have to be carefully considered and is not necessarily recommended at this time. Such correction has not yet withstood judicial test to the knowledge of the author. However, the water boundary surveyor certainly should be aware of these factors when working in areas prone to such differences.

1.3.9 Sources of Tidal Data

The National Ocean Service (NOS) of the National Oceanic and Atmospheric Administration (NOAA) of the U.S. Department of Commerce is the primary federal agency associated with tidal data. That agency and its predecessors, the U.S. Coast Survey, the U.S. Coast and Geodetic Survey, and the National Ocean Survey, have monitored tides along our nation's coastline since 1855. The data have been used historically by those agencies in connection with hydrographic surveys of the coast and for preparing predictions of the tides.

NOS maintains a network of permanent tide stations. In addition, the agency has established a large number of short-term stations in connection with many years of hydrographic surveys or in cooperative programs with various states for coastal boundary purposes. For all of these stations, NOS provides station reports containing descriptions of the bench marks associated with the stations and elevations for those bench marks relating to various tidal datum planes. That information, as well as the source data and summaries used in preparing it, is available from the agency. In addition, several of the coastal states maintain repositories of this data for the tide stations in those states.

1.4 CASE STUDIES

To illustrate typical applications of some of the theory discussed in this section, two case studies are presented involving the computation of a tidal datum. The first case illustrates the secondary determination of a tidal datum by using a year of simultaneous observation with a primary station. The second illustrates an ultimate use of tidal data with the determination of a local tidal datum by very short-term simultaneous observations and the subsequent use of the datum in locating and mapping the mean high water line.

Case Study I

The first study involves a primary tide station located on the Texas Gulf Coast and a subordinate station located about 40 miles up a nearby bayou where a tidal datum was needed. A year of simultaneous observations, from analog to digital tide gauges, was available for the stations. That was used for a computation of a tidal datum for the subordinate station.

The first step in data reduction was the selection of the high and low extremes for each tidal cycle. This was accomplished by a computer program for each month for each station. Figure 1.16 shows extremes for a typical month at the control station. Mean values for higher high water, high water, tide level, low water, and lower low water were then determined for each month by simple averaging of the appropriate extremes. Figure 1.17 illustrates a tabulation of these means for the entire year of observation.

Computations of equivalent 19-year mean values were then performed for the subordinate station by use of the Standard Method. Figure 1.18 shows the computation.

Day	High Time (h)	High Height (ft)	Low Time (h)	Low Height (ft)
1	11.1	5.72[a]		
2	10.2	5.68[a]	1.1	4.17[a]
3	11.0	5.55[a]	1.4	4.52[a]
4	12.4	5.43[a]	3.2	4.46[a]
	23.7	5.17	19.5	4.94
5	10.8	5.66[a]	3.2	4.66[a]
	23.3	5.23	18.3	5.00[a]
6	11.3	5.49[a]	3.5	5.03
			18.9	4.95[a]
7	4.1	5.84	7.0	5.69
	12.9	5.98[a]	19.0	5.17[a]
8	10.9	6.44[a]	19.2	5.07[a]
9	2.3	6.43	6.1	6.30
	9.8	6.57[a]	19.5	4.59[a]
10	11.9	6.57[a]	20.8	4.68[a]
11	8.5	6.55[a]	20.7	4.62[a]
12	5.5	6.71[a]	22.1	4.03[a]
13	6.7	6.24[a]	23.2	3.65[a]
14	7.7	5.81[a]		
15	8.3	5.70[a]	0.5	3.68[a]
	15.9	5.62	11.8	5.48
16	10.0	5.48	0.6	3.78[a]
	18.1	5.99[a]	14.8	5.35
17	10.7	5.46[a]	2.1	4.31[a]
	22.0	5.42	12.7	4.92
18	6.9	5.39[a]	1.9	4.29[a]
	19.4	4.98	16.7	4.69
			21.9	4.73
19	0.4	4.89	6.3	4.56[a]
	11.1	5.02[a]	18.1	4.35[a]
20	0.5	5.35	5.0	5.24
	10.4	5.37[a]	18.5	4.32[a]
21	2.4	5.57[a]	18.5	4.13[a]
22	3.6	5.83[a]	19.4	4.09[a]
23	4.4	5.85[a]	20.2	4.09[a]
24	6.0	5.75[a]	21.6	3.72[a]
25	5.4	5.35[a]	22.1	3.26[a]
26	8.9	5.28[a]	11.9	4.95
	14.2	5.15	22.1	3.55[a]
27	14.0	5.67[a]	22.8	4.38[a]
28	7.6	5.92[a]	23.9	4.67[a]
29	9.2	6.17[a]	23.2	4.49[a]
30	8.8	5.80[a]		

[a]Denotes higher high/lower low.

Mean Values for Month:

MHHW = 5.82 MHW = 5.71

MLW = 4.56 MLLW = 4.32

Figure 1.16 Abstract of tidal extremes at control station, June.

	Control				Subordinate			
	HHW	HW	LW	LLW	HHW	HW	LW	LLW
Jan	5.29	5.15	4.02	3.67	1.80	1.66	0.42	0.07
Feb	5.42	5.30	4.24	3.96	2.00	1.82	0.64	0.36
Mar	5.73	5.57	4.55	4.26	2.11	1.95	0.69	0.38
Apr	4.97	4.89	3.82	3.52	1.29	1.24	0.04	−0.23
May	5.46	5.37	4.19	3.98	2.23	2.11	0.66	0.48
Jun	5.82	5.71	4.56	4.32	2.52	2.37	0.97	0.72
Jul	5.68	5.58	4.45	4.34	2.33	2.19	0.74	0.64
Aug	5.24	5.21	4.20	3.87	1.68	1.70	0.59	0.29
Sep	5.68	5.60	4.58	4.27	2.08	1.98	0.78	0.49
Oct	5.64	5.56	4.52	4.21	2.07	1.98	0.76	0.48
Nov	5.71	5.68	4.55	4.21	2.15	2.08	0.76	0.49
Dec	5.48	5.47	4.19	3.92	1.86	1.92	0.54	0.36
				Annual Mean				
	5.51	5.42	4.32	4.05	2.01	1.92	0.63	0.38

Figure 1.17 Abstract of monthly means.

Known Data at Control Station:

$MHHW_c$ = 19-year mean higher high water = 5.40
MHW_c = 19-year mean high water = 5.31
MLW_c = 19-year mean low water = 4.34
$MLLW_c$ = 19-year mean lower low water = 3.99

Observed Data at Control Station:

HHW_c = annual mean higher high water = 5.51
HW_c = annual mean high water = 5.42
LW_c = annual mean low water = 4.32
LLW_c = annual mean lower low water = 4.05

Observed Data at Subordinate Station:

$HHW_s = 2.01$
$HW_s = 1.92$
$LW_s = 0.63$
$LLW_s = 0.38$

Computation of Subordinate Data:

MR_c = 19-year mean range = $MHW_c - MLW_c = 5.31 - 4.34 = 0.97$
R_c = annual mean range = $HW_c - LW_c = 5.42 - 4.32 = 1.10$
R_s = $HW_s - LW_s = 1.92 - 0.63 = 1.29$

$$\mathrm{MTL}_c = \text{19-year mean tide level} = \frac{\mathrm{MHW}_c+\mathrm{MLW}_c}{2} = \frac{5.31+4.34}{2} = 4.82$$

$$\mathrm{TL}_c = \text{annual mean tide level} = \frac{\mathrm{HW}_c+\mathrm{LW}_c}{2} = \frac{5.42+4.32}{2} = 4.87$$

$$\mathrm{TL}_s = \frac{\mathrm{HW}_s+\mathrm{LW}_s}{2} = \frac{1.92+0.63}{2} = 1.28$$

$$\mathrm{MR}_s = \frac{(\mathrm{MR}_c)(\mathrm{R}_s)}{R_c} = \frac{(0.97)(1.29)}{1.10} = 1.14 \quad (1.2a)$$

$$\mathrm{MTL}_s = \mathrm{TL}_s - (\mathrm{TL}_c - \mathrm{MTL}_c) = 1.28 - 4.87 + 4.82 = 1.23 \quad (1.3a)$$

$$\mathrm{MHW}_s = \mathrm{MTL}_s + \frac{\mathrm{MR}_s}{2} = 1.23 + \frac{1.14}{2} = 1.80 \quad (1.4)$$

$$\mathrm{MLW}_s = \mathrm{MTL}_s - \frac{\mathrm{MR}_s}{2} = 0.66 \quad (1.5)$$

$$\mathrm{MDHQ}_c = \text{19-year diurnal high water inequality} = \mathrm{MHHW}_c - \mathrm{MHW}_c = 5.40 - 5.31 = 0.09$$

$$\mathrm{MDLQ}_c = \text{19-year diurnal low water inequality} = \mathrm{MLW}_c - \mathrm{MLLW}_c = 4.34 - 3.99 = 0.35$$

$$\mathrm{DHQ}_c = \text{annual diurnal high water inequality} = \mathrm{HHW}_c - \mathrm{HW}_c = 5.51 - 5.42 = 0.09$$

$$\mathrm{DHQ}_s = \mathrm{HHW}_s - \mathrm{HW}_s = 2.01 - 1.92 = 0.09$$

$$\mathrm{DLQ}_c = \text{annual diurnal low water inequality} = \mathrm{LW}_c - \mathrm{LLW}_c = 4.32 - 4.05 = 0.27$$

$$\mathrm{DLQ}_s = \mathrm{LW}_s - \mathrm{LLW}_s = 0.63 - 0.38 = 0.25$$

$$\mathrm{MDHQ}_s = \frac{(\mathrm{DHQ}_s)(\mathrm{MDHQ}_c)}{\mathrm{DHQ}_c} = \frac{(0.09)(0.09)}{0.09} = 0.09 \quad (1.6)$$

$$\mathrm{MDLQ}_s = \frac{(\mathrm{DLQ}_s)(\mathrm{MDLQ}_c)}{\mathrm{DLQ}_c} = \frac{(0.25)(0.35)}{0.27} = 0.32 \quad (1.7)$$

$$\mathrm{MHHW}_s = \mathrm{MHW}_s + \mathrm{MDHQ}_s = 1.80 + 0.09 = 1.89 \quad (1.8)$$

$$\mathrm{MLLW}_s = \mathrm{MLW}_s - \mathrm{MDLQ}_s = 0.66 - 0.32 = 0.34 \quad (1.9)$$

Figure 1.18 Computation of tidal data by the Standard Method.

Case Study II

The second case study involves a survey of a tract of land on a barrier island off the northwest coast of Florida. The tract was located on the inshore side of the island. It is illustrated in Figure 1.19. The tract had an open shore line, for the most part, with marshy embayments at either end. A tide station existed on the island about 2.5 miles from one end of the project. During the survey, additional tide stations were established at six points within the project itself, using the existing tide station as control. The locations of these points are also illustrated in Figure 1.19.

At the control station, an analog gauge and staff were installed at the end of an old dock, very close to the original gauge location when the station was established. Two tidal bench marks were recovered at the existing station. By using the published elevation for the two bench marks, levels were run from the bench marks to the staff to determine the staff reading of mean high water. Several comparisons between the recording gauge and the staff at the station determined the gauge/staff relationship.

At the subordinate stations, simple graduated staffs, such as shown in Figure 1.20, were installed. Each staff was located in the intertidal zone, very close to the estimated line of mean high water. In the two marsh areas, this involved placement of the staffs in the marsh grass itself. Two bench marks were established in the vicinity of each subordinate staff and levels were run between the staff and the marks to preserve the datum determined by the observations.

Because an analog recorder was used at the control station, the water level was recorded continually at that point. Observations were taken at the subordinate stations at 5-minute intervals by personnel reading the staffs only for an hour or so before and after high tide. Predicted times of high water were used for deployment of personnel. An abstract and plot of readings for a typical tidal peak for both the control and subordinate station are shown in Figures 1.21 and 1.22, respectively.

By using the observed data, mean high water at the subordinate station was computed by means of the Amplitude Ratio Method. Figure 1.23 shows a typical computation. Note that the three observations have a standard deviation of 0.02 foot. A deviation such as this is typical when performing simultaneous observations at 2 or 3 miles distant from a control station in the same body of water under normal weather conditions.

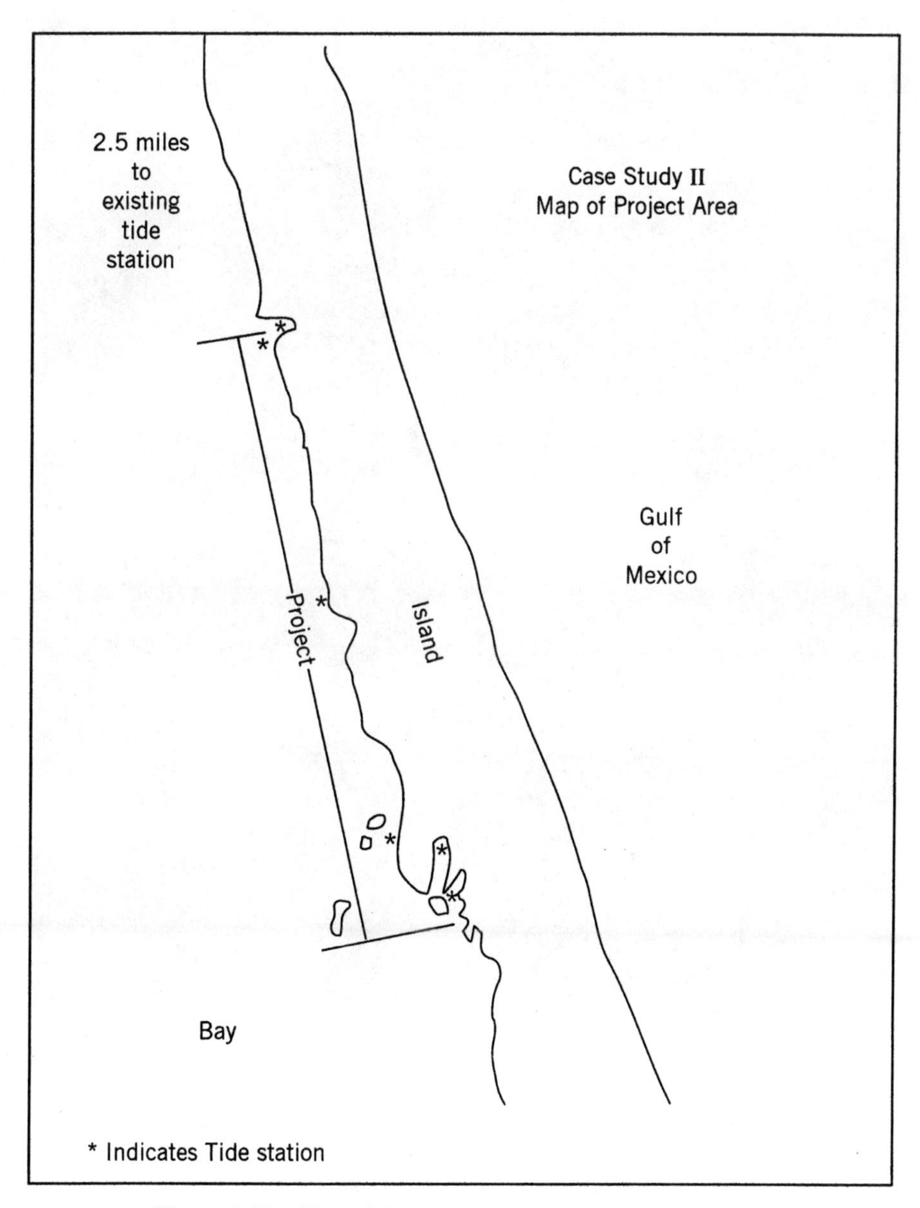

Figure 1.19 Map of the project area for Case Study II.

Figure 1.20 A tide staff.

Following similar determination of the elevation of mean high water at each of six subordinate stations, the mean high water line was identified and mapped. To accomplish this, a control-level line was run between the various subordinate stations along the project shoreline. This allowed the determination of any difference in the elevation of mean high water at consecutive stations that might require interpolation. In addition, this provided a set of temporary bench

Time Control Station		Subordinate Station 1	Time Control Station		Subordinate Station 1
(Readings in feet above staff zero)					
11:35	59.38	68.26	13:00	59.58	68.47
11:40	59.39	68.28	13:05	59.58	68.47
11:45	59.41	68.30	13:10	59.58	68.47
11:50	59.42	68.32	13:15	59.51	68.47
11:55	59.45	68.33	13:20	59.58	68.47
12:00	59.46	68.34	13:25	59.58	68.47
12:05	59.47	68.37	13:30	59.56	68.48
12:10	59.50	68.38	13:35	59:53	68.46
12:15	59.51	68.38	13:40	59.53	68.44
12:20	59.53	68.40	13:45	59.52	68.42
12:25	59.53	68.43	13:50	59.49	68.40
12:30	59.53	68.44	13:55	59.46	68.38
12:35	59.55	68.45	14:00	59.45	68.38
12:40	59.56	68.46	14:05	59.42	68.30
12:45	59.56	68.46	14:10	59.40	68.31
12:50	59.57	68.46	14:15	59.39	68.28
12:55	59.58	68.47	14:20	59.38	68.26
			20:00	57.69	
			(observed low extreme)		

Figure 1.21 Abstract of tide observations.

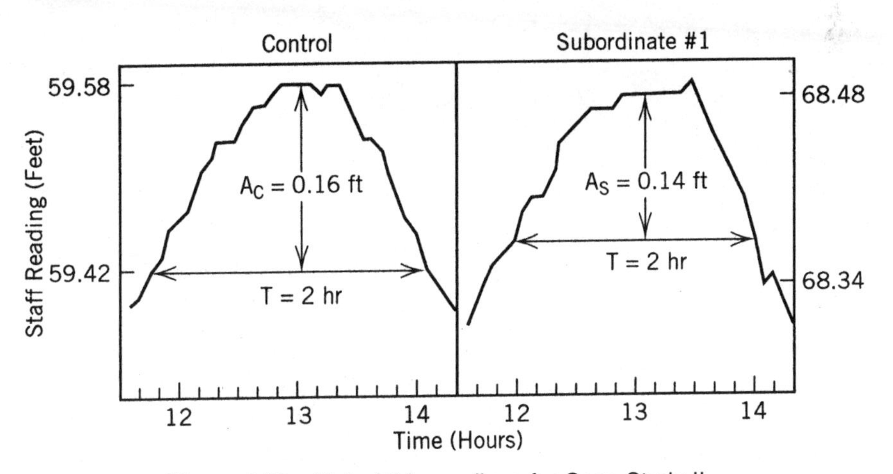

Figure 1.22 Plot of tide readings for Case Study II.

Known Data at Control Station:

MHW_c = mean high water on staff. = 59.12 ft

MR_c = mean range = 1.51 ft

$MTL_c = \text{mean tide level} = MHW_c - \frac{MR_c}{2} = 58.36 \text{ ft}$

Observed Data at Control Station (see Figure 1.22):

HW_c = observed high water on staff = 59.58 ft

LW_c = observed low water on staff = 57.69 ft

$TL_c = \text{observed half tide level} = \frac{59.58 + 57.69}{2} = 58.64 \text{ ft}$

A_c = peak height above 2-hour time interval
= 59.58 − 58.42 = 0.16 ft

$R_c = \text{observed range} = HW_c - LW_c = 59.58 - 57.69 = 1.89 \text{ ft}$

Observed Data at Subordinate Station (see Figure 1.22):

HW_s = observed high water on staff = 68.48 ft

A_s = peak height above 2-hour time interval
= 68.48 − 68.34 = 0.14 ft

Computation of Subordinate Station Data:

$$R_s = \text{observed range} = \frac{(A_s)(R_c)}{A_c} = \frac{(0.14)(1.89)}{0.16} = 1.65 \text{ ft} \quad (1.11)$$

$$MR_s = \text{mean range} = \frac{(A_s)(MR_c)}{A_c} = \frac{(0.14)(1.51)}{0.16} = 1.32 \text{ ft} \quad (1.12)$$

$$TL_s = \text{observed tide level} = HW_s - \frac{R_s}{2} = 68.48 - \frac{1.65}{2} = 67.66 \text{ ft}$$

$$MTL_s = \text{mean tide level} = TL_s - (TL_c - MTL_c) = 67.66 - (58.64 - 58.36) = 67.38 \text{ ft} \quad (1.4a)$$

$$MHW_s = MTL_s + \frac{MR_s}{2} = 67.38 + \frac{1.32}{2} = 68.04 \text{ ft} \quad (1.5)$$

Observation	*Elevation of MHW Above Staff Zero*
Observation No. 1	68.04 ft
Observation No. 2	68.04 ft
Observation No. 3	68.08 ft

Figure 1.23 Sample computation by the Amplitude Ratio Method.

marks along the coastline for subsequent identification of the mean high water line. By leveling from the temporary bench marks, points on the mean high water line were identified at frequent intervals and at each inflection point of the shoreline. As each point was identified by leveling, it was immediately located horizontally by turning angles and measuring distances from a previously established horizontal control baseline. The resulting mean high water line and the upland boundaries of the tract were plotted on rectified aerial photography.

Although procedures will vary with different locations and the purpose of the survey, the foregoing represents a typical tidal boundary survey project. The procedures, as outlined, resulted in a precise measurement of the acreage of a coastal tract, the on-the-ground location of the boundaries of the tract, and a photo-based map of the boundaries.

2

NONTIDAL SOVEREIGN/ UPLAND WATER BOUNDARIES

2.1 BACKGROUND AND HISTORY

The public trust doctrine also extends to navigable nontidal waters in many areas. This extension is not necessarily a modern innovation. As an illustration, the Roman Institutes of Justinian expressly declared rivers to be owned in common by the people in whose territory they lie (Sandars 1876).

In contrast, early English common law considered only tidal waters to be sovereign:

> That rivers not navigable (that is, freshwaters rivers of what kind so ever) do, of common right belong to the owners of the soil adjacent. But that rivers, where the tide ebbs and flows, belong to the State or public. (Hale 1666)

This distinction is perhaps because that as a small island kingdom, England has few inland waters with a capacity for public navigation.

U.S. case law has differed with English common law in this regard and has not limited sovereign waters to tidal waters. In 1876, the case of *Barney v. Keokuk* ruled that state title in navigable waters extended to inland waters as well as tidal waters with the following words:

> The confusion of navigable with tide water, found in the monuments of the common law, long prevailed in this country, notwithstanding the broad differences existing between the extent and topography of the British island and that of the American continent. . . . And since this court . . . has declared that the Great Lakes and other navigable waters of the country, are, in the strictest sense, entitled to the denomination of navigable waters, and amenable to the admiralty jurisdiction, there seems to be no sound reason for adhering to the old rule as to the proprietorship of the beds and shores of such waters. It properly belongs to the States by their inherent sovereignty. . . .

Thus, it may be generally stated that in the United States with few exceptions, the several states, in their sovereign capacity, hold title to the beds under navigable waters regardless of whether the waters are tidally affected.

2.2 BOUNDARY DEFINITIONS IN NONTIDAL WATERS

Legal definitions of sovereign/upland boundaries in waters not affected by tides are now examined. With the lack of a predictable rising and falling of the water level found in tidal waters, obviously different definitions must apply. To distinguish from the mathematically derived boundary of tidal waters (mean high water), the boundary of nontidal waters is generally called the *ordinary high water line* or *ordinary high water mark.*

English common law offers little guidance regarding nontidal water boundaries. As has been previously discussed, during the period when tidal boundaries were being defined in

England, only tidal waters were considered public domain. As indicated earlier, American case law recognized the topographic differences between the British Isles and the U.S. continent early on and declared the Great Lakes and other navigable nontidal waters of this country to be sovereign. This presumably created somewhat of a dilemma for water boundary determination due to the lack of repeating tidal cycles, which was the accepted basis for tidal water boundaries. To resolve this, U.S. case law adopted the physical fact test to determine the equivalent of mean high water in nontidal waters.

The leading definition (Maloney 1978) in federal case law, *Howard v. Ingersoll*, gives the following instructions for determining the boundary of such waters:

> This line is to be found by examining the bed and banks and ascertaining where the presence and action of waters are so common and usual and so long continued in all ordinary years, as to mark upon the soil of the bed a character distinct from that of the banks, in respect to vegetation, as well as in respect to the nature of the soil itself.

Most state case law conforms substantially with federal law on this subject. The Florida case of *Tilden v. Smith* illustrates this:

> High-water mark, as a line between a riparian owner and the public, is to be determined by examining the bed and banks, and ascertaining where the presence and action of the water as so common and usual, and so long continued in all ordinary years as to mark upon the soil of the bed a character distinct from that of the banks, in respect to vegetation as well as respects the nature of the soil itself. High-water mark means what its language imports—a water mark.

Traditionally, in nontidal waters, the courts have allowed the use of botanical and geological evidence, as evidenced by the preceding decisions, and disallowed the use of mathematical averaging of water levels. This is typified by the court's decision in *Kelly's Creek and N.W.R. Co. v. United States:*

> The high water mark is not to be determined by arithmetical calculation; it is a physical fact to be determined by inspection of the river bank.

Recently, however, there has been an apparent trend to place more reliability on water-level records, possibly due to the growing need for the precision, repeatability, and lack of ambiguity that results from a mathematical solution. Typical of this are two Florida cases, *U.S. v. Parker* and *U.S. v. Joder Cameron*. The court in the Cameron case found as follows:

> There is no logical reason why a fourth approach to determining the line or ordinary high water may not consist of comparing reliable water stage and elevation data. Indeed, for a body of water whose levels fluctuate considerably with changes in climate, accurate water stage and elevation data may provide the most suitable method for determining the ordinary high water mark.

2.3 TECHNIQUES FOR LOCATING NONTIDAL BOUNDARIES

Significantly different definitions of sovereign/upland boundaries exist in nontidal waters from those used in tidal waters, so significantly different techniques are used for their location. The emphasis of the location of the ordinary high water line in nontidal waters has been, traditionally, on the use of physical features rather than on mathematical averaging of water-level data.

When considering nontidal water boundaries, it is helpful to understand the general characteristics of the shoreline formation process. The margin of most water bodies tend to form a profile such as depicted in Figure 2.1. Obviously, the width and slope of the flood plain, foreshore and near shore slope will vary considerably with different water bodies. In addition, in riverine systems, there may be considerable variation due to the meandering of the river, including the existence of a natural levee at the waterward edge of the flood plain. However, the general characteristics should be similar for most water bodies.

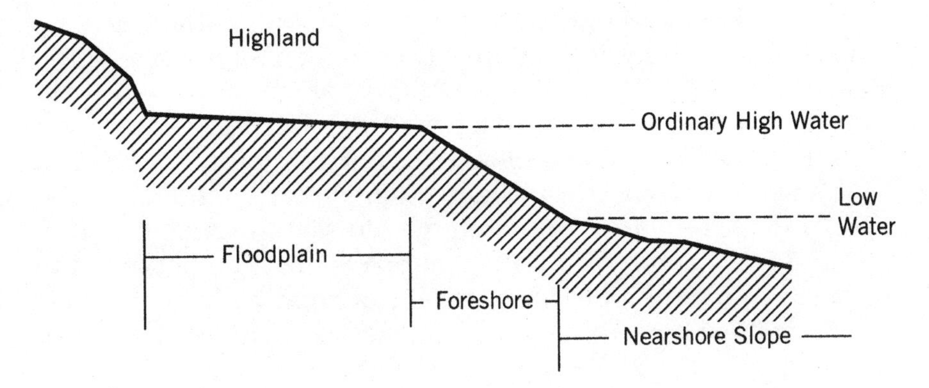

Figure 2.1 Typical shoreline configuration.

Based on various case law (see Section 6.4 of this text), the flood plain is generally not considered to be within the bed of the water body. Therefore, the objective of an ordinary high water determination is to locate the dividing line between the flood plain and the foreshore. In addressing this problem, the *Manual of Instruction for the Survey of the Public Lands of the United States* (Bureau of Land Management 1973) reads as follows:

> The most reliable indicator of mean (ordinary) high water elevation is the evidence made by the water's action at its various stages, which are generally well marked in the soil. In timbered localities, a very certain indication of the locus of the various important water levels is found in the belting of the native forest species.

The following subsections describe the various types of evidence, including that prescribed before, which might be used for such a boundary determination (Cole 1979).

2.3.1 Changes in Composition of the Soil

For lakes, one of the most repeatable indicators of the ordinary high water line may be a change in the composition of the soil. This may be evidence of the change in the organic content of the soil or the landward termination of stratified beach deposits. Stratified beach deposits occur more graphically on lake shorelines subject to beach erosion, often at the base of beach scarps. These deposits are the result of wave erosion, which tends to transport the resulting detritus away from the uplands. Generally, this transportation results in a systematic decrease in average grain size and a tendency for the particles to become more equal in size (Krumbein 1963). Therefore, a graphic difference often can be seen between the upland or parent material and the eroded material.

To make a determination of the elevation of the changes in character of the soil, it has been found that digging a narrow trench at approximately right angles to the shoreline allows a good cross-sectional view of the sedimentary and erosional features. A topographic profile of the shore along the trench should be made. Soil samples then should be taken along the profile at a few centimeters below the surface. Even when a change in soil character is not obvious, it may be advisable to take samples because laboratory analysis sometimes indicates differences that are not readily visible to the eye.

The primary information desired from these samples is a sediment particle-size analysis. There are various means of making this determination, including sieving, observation of settling velocity, microscopic examination, and so on. However, sieving is probably the most practical for sediment of the size normally found along lakes and rivers. This is a method of passing the dried sediment through a series of standard-size screens. From statistical analysis of the results of this process, two factors may be determined. These are the average grain diameter of the sediment and the degree of sorting (the extent to which the grain sizes spread on either side of the average diameter). From the previous discussion, the features for which one would look at the ordinary high water mark would be a sudden improvement in the sorting together with the occurrence of the largest average grain size.

It should be noted that rivers often form a wide flood plain due to seasonal flooding and the meandering process, so this method probably has limited application for such water bodies.

2.3.2 Geomorphological Features

Other types of geological evidence for locating the ordinary high water line are various geomorphological features such as natural levees, scarps, and beach ridges (berms). *Natural levees* are low ridges that parallel a river course. They are the highest near the river and slope gradually away from it. In larger rivers, they may be several feet high and a mile or more in width. In other rivers, however, they may be only a few inches high and a few feet wide. Levees owe their greater height near the stream channel to the cumulative effect of sudden loss in transporting power when a river overspreads its banks (Thornbury 1954). Therefore, the ordinary high water level is usually on the steep or river side and below the crest of such features. Erosional features, such as escarpments, often may be found on the river side of levees.

In Texas, the use of natural levees has been refined to very specific techniques for determining boundaries of streams (Stiles 1952). The boundaries resulting from such techniques have been endorsed by both federal and state courts in that state *(Heard v. State, Motl v. Boyd, Oklahoma v. Texas).* The first step in the Texas process is to select the "lowest qualified bank" in the area of interest. Such a bank should be an "accretion bank" (levee) as opposed to an "erosional bank" (escarpment), should have a well-defined top and bottom, should have a depression or swale on its landward side, and should be the lowest of such banks in the area. The second step is to determine the "basic point," which is the elevation halfway between the top and bottom of the lowest qualified bank. The third step is to measure the difference of elevation between the basic point and the current surface of the water at the lowest qualified bank. The boundary then may be determined by applying the same difference of elevation to the water level up and down the stream. By this method, the boundary at any point is the same difference above or below the water surface at that point as at the lowest qualified bank.

An *escarpment* or scarp is a miniature cliff cut into the shore by wave action. A *beach ridge* is a depositional feature or the wave cut slope. Beach ridges usually have a convex shape and are systematic with the apex offset landward (Knochenmus

1967). Ridges often form at various levels in a lake, but only the highest ridge is significant in boundary determination.

An interesting discussion on the use of scarps for the location of the ordinary high water mark is found in the *Manual of Instruction* (Bureau of Land Management 1973):

> Mean (ordinary) high water elevation is found at the margin of the area occupied by the water for the greater portion of each average year. At this level a definite escarpment in the soil is generally traceable, at the top of which is the true position for the meander line. A pronounced escarpment, the result of the action of storm and flood waters, is often found above the principal water level, and separated from the latter by the storm or flood beach. Another, less evident, escarpment is often found at the average low water level, especially of lakes, the lower escarpment being separated from the principal escarpment by the normal beach or shore. While these principles properly belong in the realm of geology, they should not be overlooked in the survey of a meander line.

As mentioned in the *Manual,* scarps are also found, especially in river systems, at the extremes of the flood plain. In rivers, this may be some distance from the ordinary high water mark. The more significant scarp is found in the form of undercut slopes and cut banks near the meander channel.

Geomorphological features are useful in locating the elevation of ordinary high water. They should be used with caution, however, because they can take a relatively long time to develop. If a water body is in the process of reliction or rising in elevation, there could be several sets of these features. At such time, other types of evidence are useful in resolving the ambiguity.

2.3.3 Botanical Evidence

Another type of evidence cited in various case law is the lower limit of "terrestrial plant life" *(Goose Creek Hunting Club v. United States).* The *Manual of Instruction* gives the following directions regarding the use of such evidence:

> Where native forest trees are found in abundance bordering bodies of water, those trees showing evidence of having grown under favorable site conditions will be found belted along contour lines. Certain mixed varieties common to a particular region are found only on the lands seldom, if ever, overflowed. Another group is found on the lands which are inundated only a small portion of the growing season each year, and indicate the area which should be included in the classification of the uplands. Other varieties of the native forest trees are found only within the zone of swamp and overflowed lands. All timber growth normally ceases at the margin of permanent water.

The rationale for the use of such evidence is quite clear. It has been observed and well established that many forms of plant life are distinctly related to the amount and duration of water to which they are subjected. Some of these plants have distinct preferences for water over and around themselves, or over their roots and lower parts. There are others that do not tolerate water over the soil except for short periods of time (Davis 1972). Therefore, it seems reasonable that with knowledge of the water tolerances of the plant life for a particular geographical area, patterns of such growth may offer significant assistance in the location of the ordinary high water line.

In addition to upland vegetation, species more tolerant of the presence of water also may be helpful, especially in water bodies with gently sloping shorelines. In such areas, distinct zones of vegetation may be seen, based on the water tolerance characteristics of the various species present. Correlated with other evidence, these zones may help define the location of the boundary.

Another type of botanical evidence is the lack of vegetative growth, which often may be found in a narrow zone at and slightly below the ordinary high water line. This situation is often evident on exposed coasts in lakes where there is abundant shore vegetation above and aquatic vegetation below the line. Wave action in the erosion zone tends to prevent the growth of either the upland or aquatic species. Often, one sees an unvegetated sand bottom at such places.

Unfortunately, botanical evidence often can lead to ambiguity and misinterpretation, primarily because many species of vegetation can be very adaptive and unpredictable. The following quote may add insight to this problem:

> It is important to note that it would be impractical and unrealistic to strictly apply the ordinary high water definition where the situation calls for some departure. . . . Certainly the presence or absence of vegetation is not always conclusive. The Iowa Supreme Court stated in *State v. Sorenson,* for example, that large trees may sometimes continue to grow although covered with water at their bottoms for some period. . . . This and other cases imply the converse as well. That is, even where aquatic vegetation is found some distance inland, in marshland or other poorly drained areas, for example, the finding of a realistic ordinary high water line should not be upset. (Maloney 1978)

Therefore, botanical evidence should be used cautiously and with benefit of collaborative evidence. Generalized patterns of vegetation should be used rather than reliance on a specific species.

2.3.4 Hydrological Evidence

Water-level records are also a potentially valuable class of evidence for determination of nontidal water boundaries as is the case for tidal water boundaries. Resolution of the best method for interpretation of such data, however, is still an unsettled question. The *Manual of Instruction* provides only very cryptic instruction on this topic:

> Practically all inland water bodies pass through an annual cycle of changes, between the extremes of which will be found mean (ordinary) high water.

Likewise, some case law hints at possible approaches such as the following:

> This word (ordinary high water) . . . does not mean the abnormally low level of a lake during one of a series of excessively dry years, or the abnormally high level of a lake during an exceptional wet year or series of wet years, but the average or mean level obtaining *(sic)* under fairly normal or average

weather conditions, allowing the proper range between high and low water mark in average years. *(Tilden v. Smith)*

In the past, a common use of such records has been in the form of a stage duration curve. This is made by rearranging the stage records into a cumulative distribution graph. Such a method has not met widespread acceptance, possibly due to the arbitrariness of selecting a certain percentage of inundation. Furthermore, recent evaluation of this method indicates that, depending on the water body, a wide range of percentage values may be found for the ordinary high water line as found by more traditional indicators.

One recently suggested hydrological approach (Cole 1988) uses mean high water, similar to that for tidal waters. Instead of the peaks of daily tidal cycles, this method uses the peaks of short-term variations of water level. For an averaging period, this method uses the high waters from the entire interval over which observations are available. Instead of a simple average, however, a weighted mean is determined by regression analysis to correct for any long-term changes in water level. This is accomplished by calculating the point on the least squares line (resulting from the regression analysis) that corresponds with the date of determination. Tests of this method have indicated good agreement with evidence suggested by more traditional methods.

2.4 CASE STUDIES

To illustrate typical application of the principles of this section, several examples will be presented. These will include an examination of the sites associated with two prominent judicial rulings dealing with non-tidal boundaries, a site evaluation of the methodology traditionally used by the General Land Office, and illustration of modern day determinations in a typical river and lake.

Case Study I: *Howard v. Ingersoll*

A study of this case is instructive because it is the source of the modern definition of the ordinary high water line (Maloney 1978). It represents the initial attempt by the U.S. Supreme Court in defining the line. It should be noted that the case can be confusing because it contains three, sometimes conflicting although concurring, opinions that *"do little . . . to establish a clear definition"* (Maloney 1978). However, the case represents a definite step in the evolution of the modern definition and is therefore important for a clear understanding of that definition when coupled with later, more definitive cases.

At issue in this case was the meaning of a deed call for the boundary between the states of Georgia and Alabama, which was described as running up the western bank of the Chattahoochee River. Figure 2.2 depicts a profile view of the general configuration of the river in this area based on the case opinion and recent site observations.

The main opinion of the Court defined the line as follows:

> When banks of rivers were spoken of, those boundaries were meant which contain their waters at their highest flow, and in that

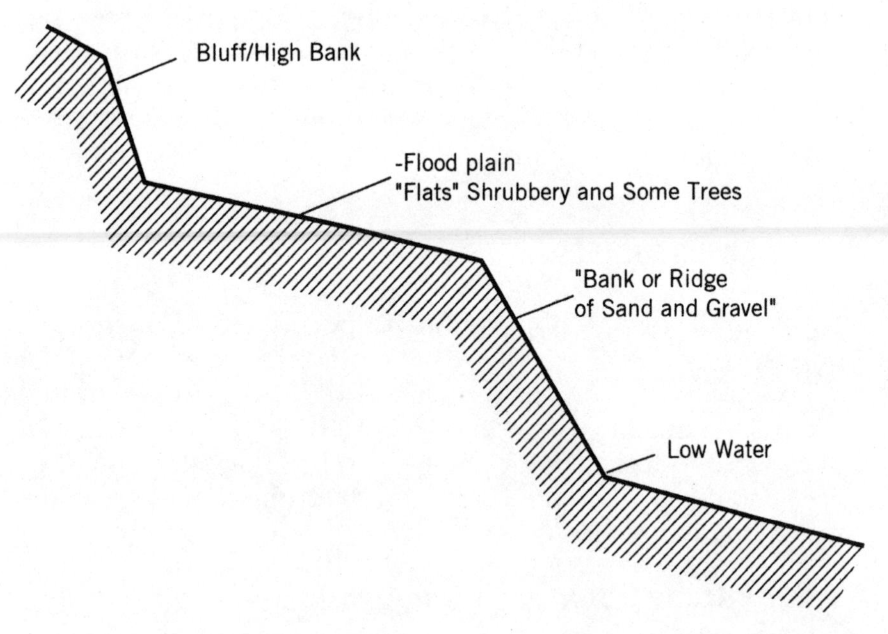

Figure 2.2 Chattahoochee River Configuration *(Howard v. Ingersoll)*.

condition, they make what is called the bed of the river . . . the outer line on the bed of a river . . . it neither takes in overflowed lands beyond the bank, nor includes swamps or low grounds liable to be overflowed, but reclaimable for meadows or agriculture. . . . The call is for the bank, the fast land which confines the water of the river in its channel or bed in its whole width, that is to be the line. The bank or the slope from the bluff or perpendicular of the bank may not be reached by the water for two thirds of the year, still the water line impressed upon the bank above the slope is the line. . . .

According to a noted water law authority *(Maloney 1978)*, this explanation "was confusing and tended towards what with hindsight we would call over-breadth . . . the main problem with this version is that it apparently contemplates drawing the line at the 'highest flow' or stage of the river." It is noted that the point of highest flow appears to conflict with the exclusion of overflowed lands from the bed of the river, which is also called for in this passage.

A concurring opinion, written by Justice Nelson and concurred to by Justice Grier, placed the line at the ordinary level as follows:

In the ordinary state of the river . . . the water covers this flat about halfway to the high bluff, extending to the base of a bank or ridge of sand or gravel: and in freshet, the water covers the flats reaching to the bluff. (The line) does not necessarily, nor as I think, reasonably, call for a line along the bluff or high bank. . . . The bank enclosing the flow of water, when at its ordinary and usual stage, is equally within the description; and the limit within this bank, on each side, is more emphatically the bed of the river, than that embraced within the more elevated banks when the river is at flood. . . . In our judgment, the true boundary line . . . is the line marked by the permanent bed of the river by the flow of the water at its usual and accustomed stage. . . .

A second concurring opinion by Justice Curtis added greater definition to the boundary by defining it "by reference to several ascertainable physical characteristics of the bank" (Maloney 1978) as follows:

(The) line is to be found by examining the bed and banks, and ascertaining where the presence and action of water are so common and usual and so long continued in all ordinary years, as to mark upon the soil of the bed a character distinct from that of the banks, in respect to vegetation, as well as in respect to the nature of the soil itself. Whether this line . . . will be found above or below, or at a middle stage of water, must depend upon the character of the stream. . . . But in all cases the bed of a river is a natural object . . . the banks being fast land, on which vegetation, appropriate to such land in the particular locality, grows wherever the bank is not too steep to permit such growth, and the bed being soil of a different character and having no vegetation, or only such as exists when commonly submerged in water.

Later cases, as previously suggested, appear to have followed the latter two opinions as opposed to the primary opinion in this case.

Case Study II: *Tilden v. Smith*

The case of *Tilden v. Smith* represents a typical state court decision dealing with ordinary high water. This case was decided by the Supreme Court of Florida in 1927. It involved a dispute over the rights of riparian owners of Lake Johns in Orange County, Florida. The actual question before the court was whether one riparian owner had the right to lower the level of the lake, when it was in an overflowed state, by use of a drainage well. Although the location of the ordinary high water line was not the primary question addressed by this case, examination of this opinion, coupled with examination of the lake involved, allows insight into what the court considered to be the ordinary high water line and especially its treatment of the flood plain.

Lake Johns, unlike many Florida Lakes, is not fed by underground springs, but by rainfall and runoff. Therefore, there may be considerable variation in water level from year to year due to meteorological cycles. The court took note of this large variation but stated that *"nevertheless the character of the vegetation and trees around the lake gave some evidence of an average or ordinary high water mark. . . ."* At the time of the litigation, the lake was at an abnormally high stage. A golf course of the West Orange Country Club, which was located on the flood plain, was submerged and nearby cottages made nonhabitable by the high water level.

The deep well drilled for drainage purposes was, according to the evidence introduced, situated "at a point higher than the level of the lake within its natural boundaries." The disputed well, in addition to a second well drilled at approximately the same time, has been recently recovered and leveled to from nearby vertical control. This resulted in elevations for the top of the casings of these two wells of 92.6 and 92.8 feet with National Geodetic Vertical Datum (NGVD) (Gentry 1989).

An examination of the shore of the lake provides evidence of the ordinary high water line of both geomorphological and botanical nature. By looking at the geomorphological evidence, a pronounced escarpment exists at the upper limit of the flood plain at an elevation of approximately 98.5 feet. In addition, an escarpment is found at 90 feet and a low water escarpment is located at 88 feet (Gentry 1989). This profile matches the classic profile described in the quote from the *Manual of Instruction* provided in Section 2.2.2. Based on those instructions, the ordinary high water line would be along the middle escarpment, which is slightly below the elevation of the wells as described by the court opinion.

The flood plain of Lake Johns is relatively devoid of upland trees, apparently because of the nature of the water-level fluctuations in the lake. This is shown in Figure 2.3, where the zoning of shorter-term vegetation may be observed. The apparent ordinary high water level, based on vegetation, is at the waterward edge of scattered shrublike vegetation. There is also a distinct change in the grasses at that point with more water-tolerant species emerging. Leveling indicates that this occurs at an elevation of 90 feet (Gentry 1989), which is consistent with the geomorphological evidence as well as the court's description of the location of the line.

Figure 2.3 Floodplain at Lake Johns *(Tilden v. Smith).*

Case Study III: Lake Butler GLO Survey (Gentry 1989)

Lake Butler is also located in Orange County in Central Florida. However, it is spring-fed and therefore not as volatile in elevation as Lake Johns. Lake Butler was selected as a case study because a portion of its shoreline was resurveyed in the twentieth century by government surveyors. Therefore, it provides a location for comparison of current evidence with a relatively recent survey by federal surveyors, thus allowing observation of their practice.

The lake was originally surveyed by the General Land Office in 1848. However, apparent omissions of land in that survey led to a resurvey in 1923 by U.S. Cadastral Engineer Arthur Brown. Figures 2.4 and 2.5 represent a copy of the plat for the resurvey and a current photograph of the area at the same scale as the plat. Note that on the plat, both the previous meander line, which is the landward line, and the resurveyed meander line are depicted.

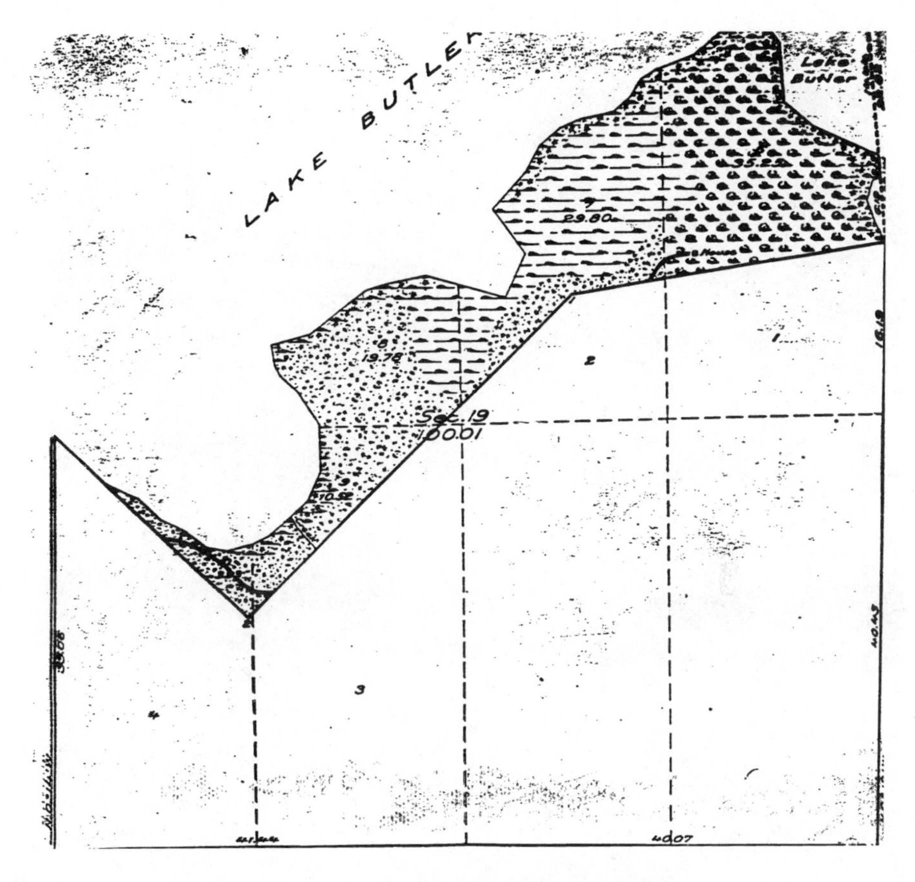

Figure 2.4 Resurvey plat of Lake Butler shoreline.

The meander line, as surveyed by Brown, has been retraced to allow observation of the meander points on the current shore of the lake (Gentry 1989). Through much of the surveyed area, the current ordinary high water line is quite apparent. Brown's notes for this area read as follows: "Cor. set at mean high water mark . . ." and "along well defined shoreline with rolling timber to right of meander line, cypress timber along shore and extending out into lake at places."

In this area today, there is a slight berm or levee with an escarpment on its waterward side. Upland vegetation is eroded away below that point with the exception of occasional buttressed cypress trees. Organic soil is found upland

Figure 2.5 Aerial photograph of Lake Butler shoreline.

of that line and a strip of bare sand bottom is seen immediately waterward of the line. Waterward of the bare strip, aquatic vegetation is dense. The retraced meander points generally fall on or very near to this line. In one area, there has been apparent erosion during the intervening years, and the meander points are about 20 feet waterward of the current, ordinary, high water line in that area.

The most instructive portion of this case study is the surveyor's treatment of the two swamp areas bordering the lake. One of these areas is in Government Lot 7 and the other on the quarter line between Government Lots 7 and 8. The instructions under which the resurvey was conducted (General Land Office 1902) state that "Lands bordered by waters are to be meandered at mean high water mark." They further instruct that "meander lines will not be established at the segregation line between dry and swamp or overflowed land, but at the ordinary high water marks of the actual margin of the rivers or lakes on which such swamps or overflowed lands border." Brown was faithful to these instructions and placed his meander line between the swamp land and the lake bed. His notes for this area read as follows: "to beginning of strip of swamp land, 1 chain in depth bet. meander line and upland." In his general description of the area, he elaborated on this placement of the line as follows:

The extension area in sec. 19 is about equally divided between swamp area and upland area. About half of the swamp area is covered with cypress, gum and other swamp timber, and the remainder is grown up with dense saw grass. Drainage operations have lowered the usual level of Lake Butler one or two feet but the lake was unusually high at the time this survey was made, due to unusually heavy rains of the preceding fall and winter. Old settlers declared the lake level to be as high, or practically as high, at the time of this survey as it used to usually stand before drainage operations ever started. However the determining factor in deciding that the swamp land portion of this extension area was land in place at the time of the original survey and at the time when Florida was admitted to the Union is the very definite old shoreline extending across most of the front of the swamp land area. Timber growths on this area easily antedate the time of the original survey, or the date when Florida was admitted into the Union.

Along that line today, definite changes in vegetative species may be noted. Scrubby plants begin immediately upland of the line. Wetland species may be seen in the swamp, but not the same types of aquatic vegetation found in the bed of the lake. Such areas should have been and were excluded according to the prevailing *Manual of Instruction* at the time of the resurvey as well as the currently prevailing manual for government surveyors.

Case Study IV: Typical River

This case study illustrates a typical ordinary determination on a river. In the river selected, there is an abundance of both geomorphological and botanical evidence. Figure 2.6 provides a generalized cross-section of the river bank. At the time of the field study, the river was at a relatively low stage. As shown in Figure 2.6, three escarpments were observed. The lowest was slight and located just above the water level at the time of the study. The middle escarpment (which is being pointed to in the figure) was somewhat more prominent and contained numerous matted roots and marked the lower limit of transitional vegetation. The highest escarpment was covered, above and slightly below, by various upland vegetation including saw palmetto. Upland trees, such as live oak and citrus, grow down to slightly above (vertically) the upper escarpment. During times of seasonal flooding, the water level has gone well above all of the escarpments. Based on these indicators, it appears that the ordinary high water line should be located along the middle escarpment.

In addition to studying the geomorphological and botanical evidence, an analysis was performed of existing hydrological evidence. Approximately 26 years of water-level gauging by the U.S. Geological Survey was available for a site near the study location. By using the newly proposed method described in Section 2.3.4, an average of all the high waters occurring in each year was computed and a

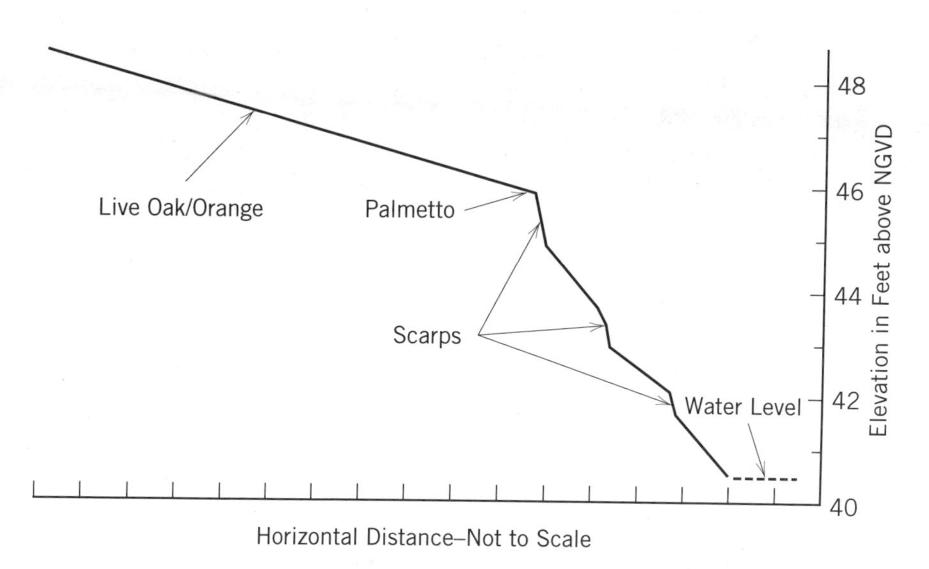

Figure 2.6 Ordinary high water study of a typical river.

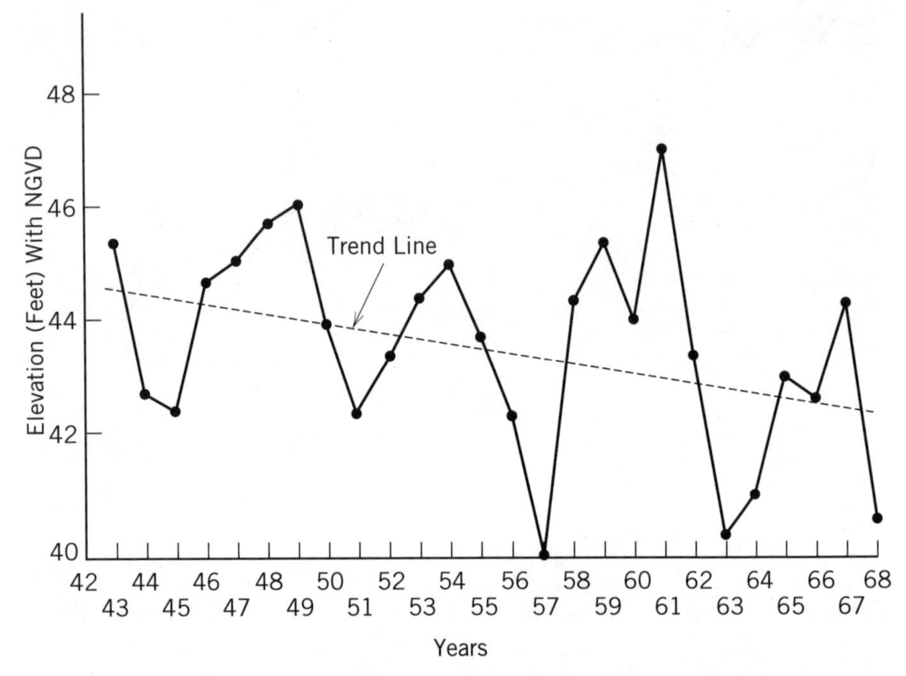

Figure 2.7 Plot of annual mean high waters of a typical river.

least squares trend line determined for the annual mean high waters. The result is shown in Figure 2.7. Using the hydrological data in this manner resulted in a computed mean high water of 42.8 feet at the end of the gauge observations, which provides excellent correlation with the middle escarpment indicated by the geomorphological and botanical evidence.

Case Study V: Typical Lake

This case study illustrates a typical ordinary high water determination on a lake. As with the river, the study was performed at a time of relatively low water to allow observation of all shoreline features. As shown in Figure 2.8, a number of escarpments were noted. One large extreme high water escarpment was located well within the timber growth surrounding the lake. Sparse timber growth continued down to a point slightly below the second highest escarpment.

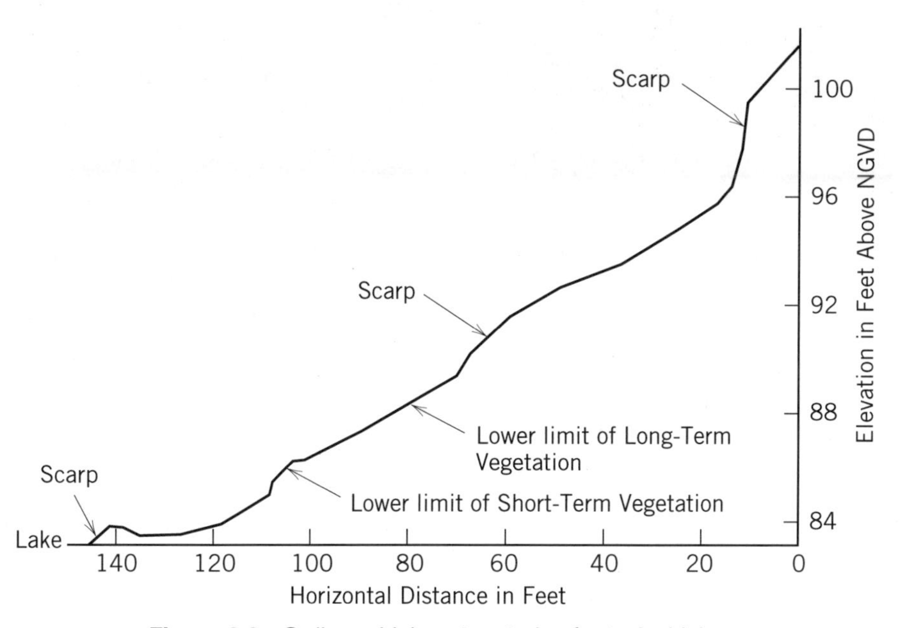

Figure 2.8 Ordinary high water study of a typical lake.

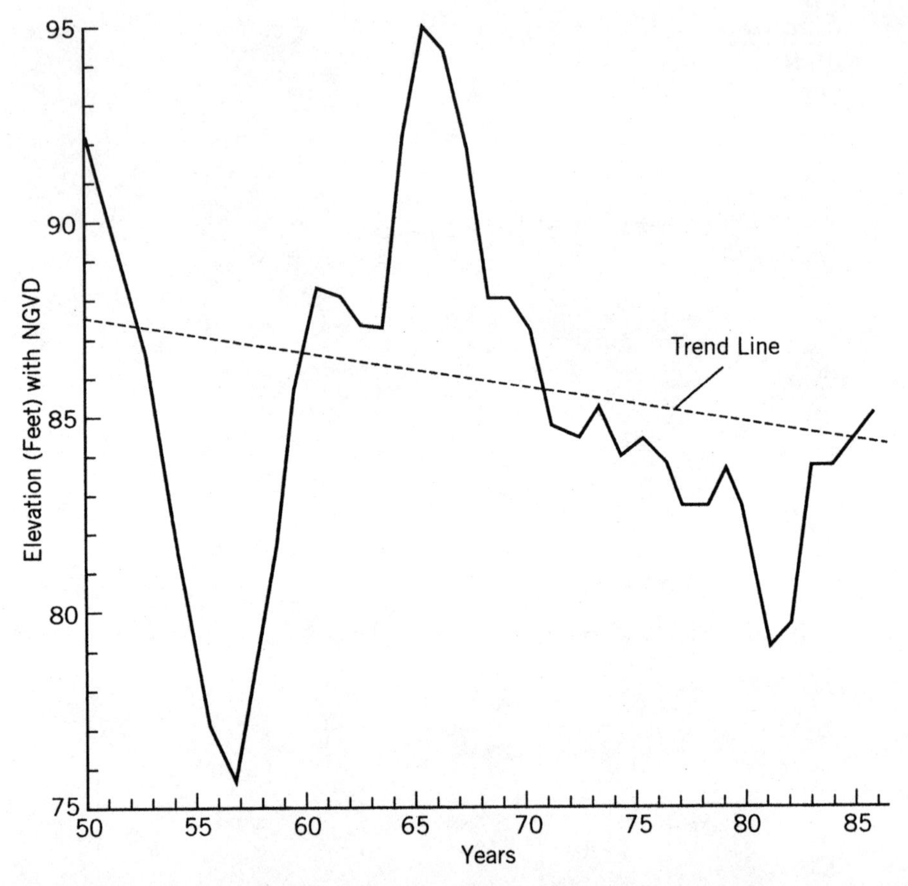

Figure 2.9 Plot of annual mean high waters of a typical lake.

Shorter-term upland vegetation, along with soils containing high organic content, continued down to a third, less prominent, escarpment. Finally, sandy soil with sparse grass was noted continuing on down to a low water escarpment near the water level at the time of the study. Based on these indicators, it appears that the ordinary high water line was located at the third highest escarpment.

About 35 years of water-stage gauging had been conducted by the U.S. Geological Survey in this lake, so the resulting data from those observations were analyzed in a method similar to that used in the river example. A plot of the resulting annual mean high waters is provided in Figure 2.9. From the figure, the least squares line through the annual mean high waters indicated that the water level had been failing at a rate of 0.09 foot per year. A mean high water was determined by the trend line for the time of the study as being 84.5 feet with NGVD. This value agrees with the third highest escarpment and tends to collaborate the other evidence.

3

RIPARIAN RIGHTS RELATED TO SOVEREIGN/ UPLAND BOUNDARIES

When dealing with riparian tracts, it is frequently necessary to delineate limits of exclusive riparian rights associated with upland tracts. Increasingly, such information is being requested in association with applications for governmental permits for use of such areas. This section outlines the general guidelines for determining such division lines.

Riparian rights are rights relative to the use of water accruing to owners of upland bordering on water bodies. In the true sense of the word, "riparian" rights (derived from the Latin *ripa,* a river bank) apply only to lands bordering rivers and streams; with "littoral" rights (derived from the Latin *litus,* the seashore) more properly applying to oceanfront lands. Because of the general use of the word, however, "riparian" is used in this discussion as including all types of water bodies.

Determination of the division line between the riparian rights area of adjacent land owners often can be a challenging study. This is primarily because the method for determining the division line may vary significantly with the situation. This is emphasized in various court opinions on this subject *(Hayes v. Bowman; Johnson v. McCowen),* which have stated that "No geometric theorem can be formulated to govern all cases." Although this is true, there are some general rules of procedure. It is emphasized, however, that such rules do vary considerably with type and size of the water body, shape of the shoreline, and other parameters. In addition, there may be considerable variation in rules from state to state. A large number of case cites from various states have been provided to assist in that regard.

It is a well-settled question in the United States that the laws of each state controls the apportionment of such rights, as opposed to federal law *(Hardin v. Jordan).* Another basic premise when dealing with such rights is that the length of the affected tract's water frontage, not its total acreage or upland lot lines, generally controls the apportionment of riparian rights.

3.1 DIVISION LINES FOR RIGHTS WITHIN ADJACENT WATERS

Consideration is first given to riparian rights to the use of adjacent waters. These include numerous items such as access to the navigation channel, the right to construct docks, filling of submerged land, and view. Many of these common law rights are subject to regulation and even total elimination by state law, so there may be variations between states in which rights apply.

To find the dividing line between exclusive areas for such rights, a point of departure must first be determined. It appears to be well settled that such points of departure should be at the intersection of the upland boundary and the actual shoreline as opposed to the meander line of government surveys *(Menasha Wooden Ware Co. v. Lawson).* It is noted that when the rights to newly formed uplands are at issue, it is necessary to use the intersection of the upland boundary and the shoreline at the time of subdivision. In such an instance, the government meander line may be the best evidence of that location and therefore may be used. Thus, in most states, the point of departure would be at the intersection of the upland boundary and the mean high water line in tidal waters or the ordinary high water line in nontidal waters. This would obviously vary in those states recognizing mean low water as the boundary of sovereign submerged lands.

Once the point of departure is established, the question of the proper direction of the dividing line must be addressed. It has been generally held that the proper procedure for this does not involve extension of the upland boundary without change of direction. However, there have been cases that have held that such an approach is appropriate where the upland boundary strikes the shoreline at approximate right angles *(Bond v. Wool, NC)*. In addition, some cases *(Lattig v. Scott, Id.; McCamon v. Stagg, Kan.; Rector v. United States)* suggest that in public land states where the land boundary is described by government sections or aliquot parts of sections, there is justification for using extension of upland boundaries where section lines continue on the same course across a river. Generally, however, the direction of the upland boundaries are not controlling for division lines for riparian rights.

3.1.1 Rivers

The method for determining the proper direction for projection of the division lines appears to vary with the nature of the water body. In rivers and streams, which are significantly narrow to determine a "thread" or center of stream, the prevailing procedure is to project the line in a direction perpendicular to the thread of the stream. Typical case law suggesting this approach is as follows (Foster 1959): Massachusetts *(Knight v. Wilder)*, Michigan *(Campau Realty Co. v. Detroit; Bay City Gaslight Co. v. Industrial Works; Clark v. Campau)*; Nebraska *(Application of Central Nebraska Public Power & Irrigation District)*; New York *(Calkins v. Hart; U.S. v. Ruggles)*; Pennsylvania *(Wood v. Appal)*; Oklahoma *(Rector v. United States)*; Oregon *(Montgomery v. Shaver)*; Wisconsin *(Farris v. Bentley; Menasha Wooden Ware Co. v. Lawson; Superior v. Northwestern Fuel Co.)*

The "thread" of a stream is usually defined as a line equidistant from the two banks of the stream. Therefore, a projection perpendicular to the thread requires the mapping of the shoreline (mean high water line or ordinary high water line in most states) on both sides of the stream. In many cases, the use of photogrammetry is helpful for this process. Once the shoreline is identified, a median line may be constructed to approximate the thread (see Section 9.2). Where the geometric center of the stream is considerably removed from the thalweg, or deepest part (channel) of the stream, there may be justification for using the thalweg as the center. Location of the thalweg requires the use of hydrographic soundings.

In wide rivers where the thread would not be readily discernible (also in large bays or oceans), case law suggests using lines run perpendicular to various other baselines such as the shoreline if it is relatively straight, the bulkhead or pier line, or the channel line. Typical cases suggesting this approach are as follows (Foster 1959): California (perpendicular to the shoreline: *Fraiser's Million Dollar Pier Co. v. Ocean Park Pier Co.);* Florida (in the direction of the channel: *Hayes v. Bowman*); Louisiana (perpendicular to the shoreline: *Municipality No. 2 v. Municipality No. 1);* Maryland (perpendicular to the shoreline: *Baltimore v. Baltimore & PSB Co.*); New Jersey (perpendicular to the shoreline: *Bradley v. McPherson; Delaware L&WR Co. v. Hunnan; Manufacturers' Land & Improvement Co. v. Board of Commerce & Navigation*); New York (perpendicular to the shoreline: *People ex rel Cornwall v. Woodruff*); North Carolina (perpendicular to the shoreline: *Oneal v. Rollinson*); Ohio (perpendicular to the shoreline: *Ludwig v. Overly*); Oregon (perpendicular to the government pierhead line: *Columbia Land Co. v. Van Dusen Investment Co.*); Rhode Island (perpendicular to the harbor line: (Aborn v. Smith; *Manchester v. Point St. Iron Works; Tabor v. Hall*); South Carolina (perpendicular to the shoreline: *McCullough v. Wall*); Virginia (perpendicular to the shoreline: *Lambert's Point Co. v. Norfolk & W. R. Co.*).

When the shoreline in question is curving or irregular, such as in coves, some courts have held that the proper procedure is to give each riparian tract a proportionate share based on measurement of the shoreline in question as well as measurement of some outer line, such as a channel line or harbor line

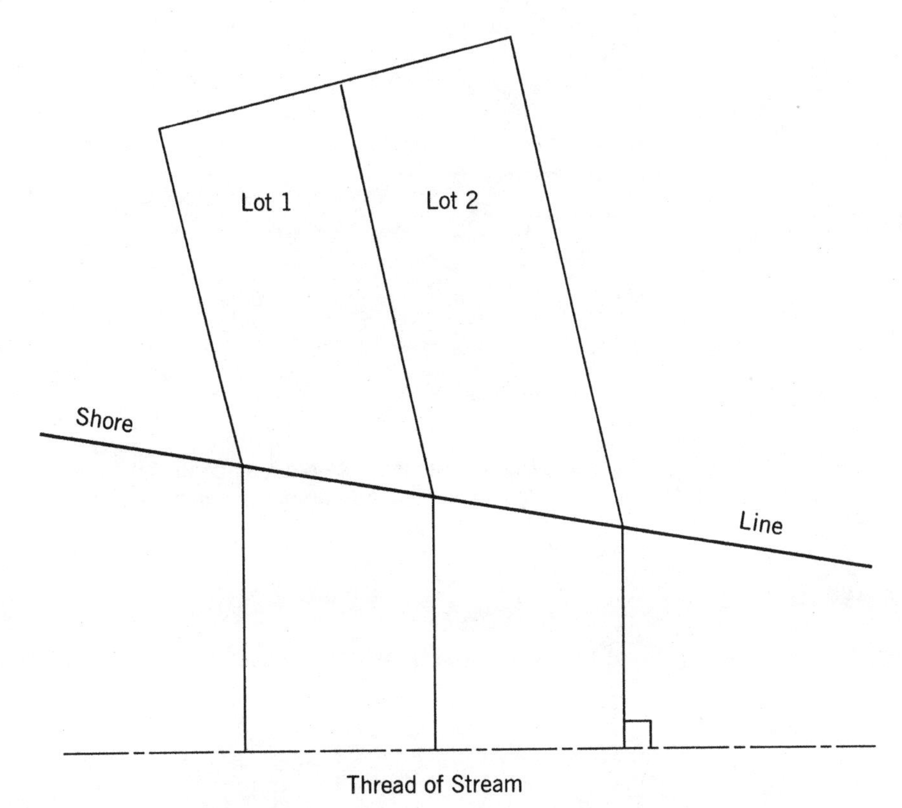

Figure 3.1 Division line perpendicular to the thread of stream.

(or new shoreline where newly formed uplands are being apportioned). Cases illustrating this approach are as follows (Foster 1959): Massachusetts *(Deerfield v. Arms);* Maryland *(Baltimore v. Baltimore & PSB Co.);* Virginia *(Cordovana v. Vipond; Groner v. Foster; Rice v. Standard Products Co.; Waverly Water Front & Improvement Co. v. White.).*

Figure 3.1 shows a typical situation illustrating division lines run perpendicular to the thread of a stream. Figure 3.2 shows a typical case dealing with a wide river where the court held that lateral lines run normal to the channel line would be the proper division lines. Figures 3.3 and 3.4 illustrate cases where the courts have indicated that proportionate measurement is the correct approach.

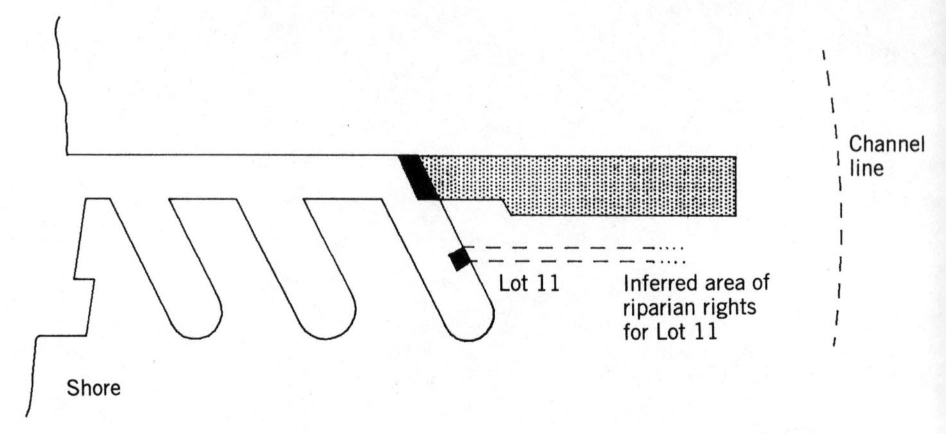

Figure 3.2 *Ruling:* Proposed filling of hatched area did not infringe on riparian rights of Lot 11, which lie "as near as practical in the direction of the channel" (from *Hayes v. Bowman*).

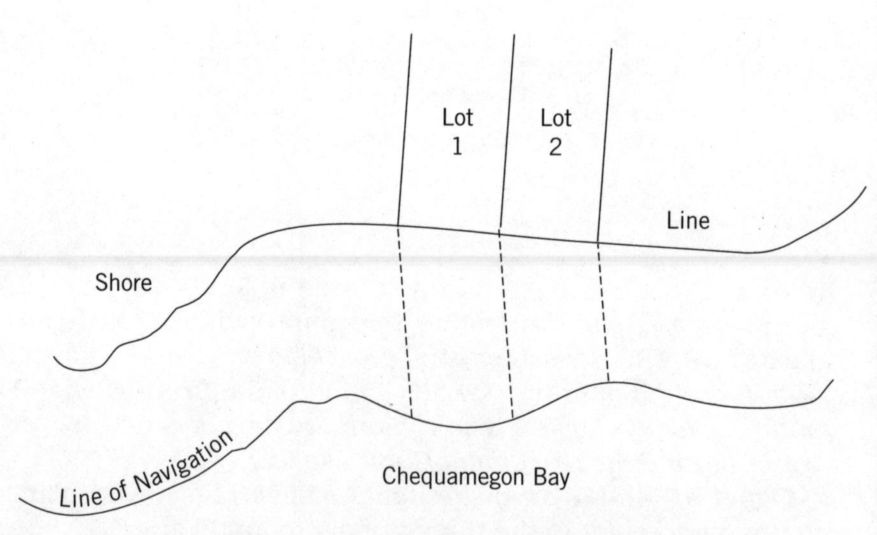

Figure 3.3 *Ruling:* "Measure the whole shoreline of the cove or bay and the line of navigable water on front of the same and apportion the latter among the owners according to the length of their respective holdings on the shoreline, drawing straight lines between the corresponding points of division on the two lines" (from *Northern Pine Land Co. v. Bigelow*).

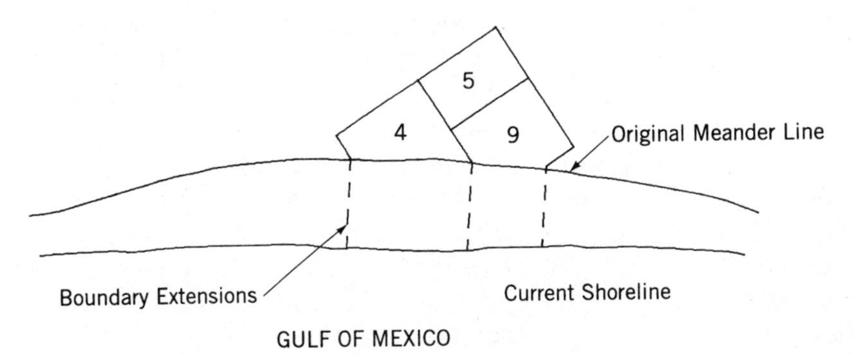

Figure 3.4 *Ruling:* The boundary should be extended "so as to provide each lot with the same relative amount of shoreline as it had on the meander line in the original survey" (from *Bliss v. Kinsey*).

3.1.2 Lakes

Where riparian rights on lakes are involved, entirely different approaches often are required because a closed figure is involved. For example, in round lakes, most cases have suggested side lines drawn to a center point, creating pie-slice shaped areas. Typical cases suggesting that approach are as follows (Sullivan 1982): Idaho *(Ulbright v. Basington),* Minnesota *(Hanson v. Rice; Markusen v. Mortensen).*

For long lakes, the courts have sometimes held that an approach similar to that used for rivers be used with a thread or center line being constructed and lines projected out perpendicular to the center line. (See Section 11.3.) Cases suggesting this approach are as follows (Sullivan 1982): United States *(Hardin v. Jordan);* Minnesota *(Rooney v. Stearns County Board; Scheifert v. Briegal);* New York *(Calkins v. Hart; Mix v. Trice);* South Dakota *(Korterud v. Darterud).*

Another approach for long lakes, especially when the lake is large and the shoreline is straight, is to use division lines run perpendicular to the shoreline. Cases suggesting this approach are as follows (Sullivan 1982): Idaho *(Driesbach v. Lynch; Hilleary v. Meyer):* Indiana *(Shedd v. American Maise Products Co.);* Washington *(Hefferline v. Langkow; Seattle Factory Sites Co. v. Saulsberry);* Wisconsin *(Northern Pine Land Co. v. Bigelow; Thomas v. Ashland).*

A common approach for irregular shaped lakes, especially when they are large, is the proportionate division of the line of navigable water, similar to that suggested for irregular shorelines in rivers. Cases suggesting this approach are as follows (Sullivan 1982): Idaho *(Driesbach v. Lynch; Hilleary v. Meyer; Randall v. Ganz)*; Michigan *(Blodgett & Davis Lumber Co. v. Peters; Stuart v. Greanyea)*; Ohio *(American Steel & Wire Co. v. Cleveland Electric Illuminating Co.)*; Washington *(Hefferline v. Langkow; Seattle Factory Sites Co. v. Saulsberry)*; Wisconsin *(Northern Pine Land Co. v. Biglow; South Shore Lumber Co. v. C. C. Thompson Lumber Co.; Thomas v. Ashland)*.

3.1.3 General Rules

By reviewing the preceding, there are four basic cases with each case having its own rules. These include narrow streams, wide rivers, lakes, and irregular shorelines. With narrow streams, the general rule is to use a division line perpendicular to the thread of the stream. With wide rivers (and elongated lakes), the general rule is to use a line run perpendicular to the shoreline, center line, or to some other baseline such as a bulkhead line. For most lakes, the general rule is to use lines drawn to a common center point or center line. For irregular shorelines, proportional division of the center line based on the length of the shoreline is the general rule.

3.2 RIGHTS OF UPLAND OWNERS TO NEWLY FORMED LAND

These rights deal with newly formed land that caused changes in the shoreline. Such changes include those caused by withdrawal of the water *(reliction)* and those caused by material being deposited along the shoreline by the water *(accretion)*.

The general rule is that the upland owner gains title to new upland created by reliction and accretion and loses title to land submerged by rises in water level or lost by erosion. However, when such changes are not gradual or imperceptible, or when such changes are artificially induced, the general rule may not apply. When such changes occur suddenly, such as during storms, this is called *avulsion* and it is generally held that title does not change with such shoreline changes. Likewise, it is usually held that shoreline changes resulting from man-made actions, such as those associated with dredging or groins, do not change title if the upland owner or a predecessor in title caused the changes. An illustration of this occurred in Florida where case law generally holds that artificial accretion caused by the upland owner remains the property of the sovereign *(McDowell v. Trustees of the Internal Improvement Trust Fund of State of Florida)*. However, case law in that state also holds that artificial accretion caused by third parties accrue to the upland owner *(Board of Trustees v. Madeira Beach Nominee Inc.; Board of Trustees v. Sand Key Associates Ltd.)*.

When changes occur along a shoreline that do result in a title change, the division line across the new lands between the adjacent riparian owners may have to be determined. Similar procedures to those for riparian rights over water are generally used, although some jurisdictions have distinguished between the two applications.

4

HISTORIC BOUNDARY LOCATION

4.1 SHORELINE CHANGES

Shorelines are dynamic in nature, and their locations may change considerably with time. Occasionally, the need arises to back up time and locate an earlier (historic) shoreline position. Examples of this occur when it is necessary to determine the quantity of land added or removed for dredge and fill regulation enforcement, for division of accreted areas, for erosion studies, or for purposes of reclaiming land lost due to avulsion. Regardless of the reason, this task presents a unique challenge to the surveyor. This is especially true because it is sometimes necessary to locate the shoreline at its last natural position prior to artificial alteration. Separation of natural and artificial changes can be an especially difficult task.

It is often helpful to consider such changes as being in two categories. These categories consist of changes associated with the lowering (reliction) or rise of water level and those associated with changes in land form. The latter category includes erosion (gradual wearing away of the land due to water and wind action) and accretion (gradual building up of land). In addition, that category includes more sudden or avulsive changes in land form such as those caused by a specific storm or by dredging or filling. Such classification may help to make the correct legal interpretation for the actual boundary location as well as in selecting the best approach for locating the historic shoreline location.

4.2 TECHNIQUES FOR LOCATING HISTORIC SHORELINES

Techniques for locating historic shorelines vary considerably with the category of shoreline change experienced. For example, if the shoreline has changed due to a lowering or rising water level without appreciable land form change, the use of water-level records is the best approach (where such records are available). In such a case, the location of the historic shoreline is a matter of analysis of the water-level records and location of the contour of the appropriate average water level as of the prechange date.

When the shoreline changes are those associated with a change in land form or when water-level records are not available, other approaches must be considered. The best course in such cases is usually the use of graphic evidence describing the land prior to the shoreline change. Such evidence might include surveys, maps, and aerial photographs. The most definitive of these is a survey plat or map that specifically delineates the desired water boundary. To find such a map of a specific area, made at a specific time, and to the precision desired, however, is often a difficult task. For periods after the 1930s, one most often finds repetitive aerial photography of a given area to be the best evidence. When photography is used, additional steps are required. These are the interpretation of the photography to identify the water boundary and the rectification of the photo imagery to a current horizontal datum.

An example of the use of historic photography is provided by Figures 4.1 and 4.2. These photographs illustrate a riverine marsh area where fill was allegedly placed waterward of the mean high water line. Figure 4.1 shows the area prior to the placement of the fill and Figure 4.2 shows the current configuration. For purposes of litigation, it was desired to approximate the location of the mean high water line across the filled area prior to filling. This was accomplished by performing a survey of the current mean high water line in the undisturbed areas on either side of the filled area. Photoidentifiable features were located during the survey to allow plotting of the surveyed mean high water line onto the historic photograph. The location of the mean high water line was then interpolated across the filled area by use of features on the photo image

Figure 4.1 Predevelopment configuration (shoreline data from the files of the Florida Department of Natural Resources).

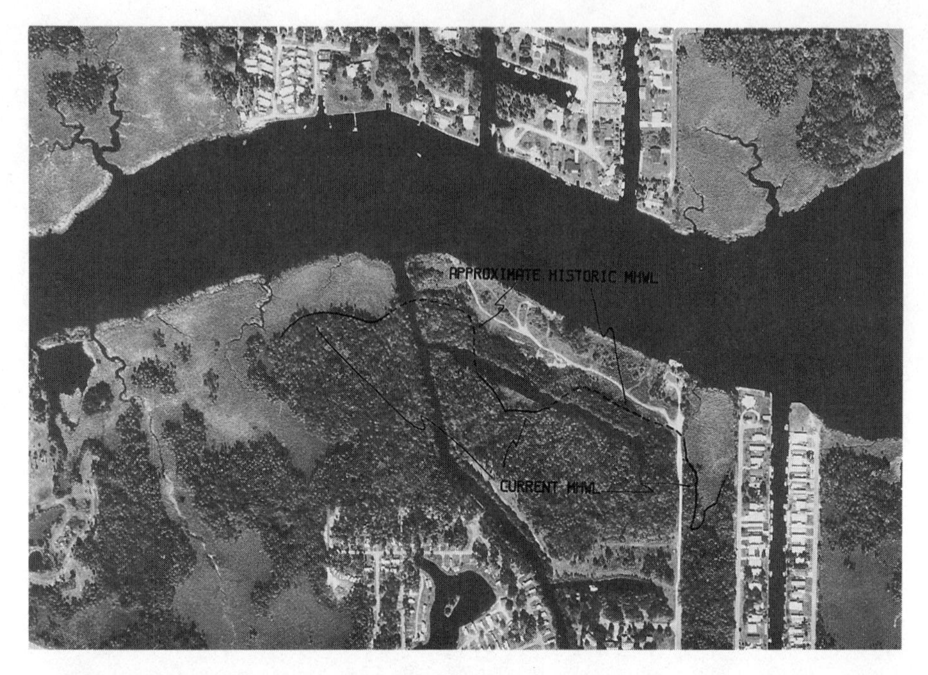

Figure 4.2 Postdevelopment configuration (shoreline data from the files of the Florida Department of Natural Resources).

using the line on either side as ground truthing. A stereo plotter is especially useful for such work because it allows the following of a contour as well as examination of the magnified image. Figures 4.1 and 4.2 show the resulting surveyed and interpolated lines for this case.

Depending on the site and available maps and photography, other evidence also may be considered for such determinations. This might include the testimony of eye witnesses, soil analysis, and vegetation analysis.

4.3 SOURCES OF INFORMATION FOR HISTORIC SHORELINES

Considerable information is available for locating historic shorelines. Some of the most common sources are detailed in the following sections.

4.3.1 Shoreline Topographic Maps

For major coastal bodies of water, some of the better sources of historic shoreline locations are shoreline topographic maps produced by the U.S. Coast Survey (later named the Coast and Geodetic Survey and presently the National Ocean Service). These maps, commonly called "T" sheets, are the original field survey manuscripts prepared for use in compiling nautical charts. As with all such data, the primary purpose of these maps must be considered when using them. Because they were produced for the primary purpose of producing nautical charts, the scale and detail may not be the same as if they were produced primarily for a water boundary survey. Therefore, the inherent limitations must be appreciated. Nevertheless, the detail on some of these maps is quite astounding. Figure 4.3 shows a typical "T" sheet.

The first of such maps was produced in 1834. Most of the coastline for the conterminous United States was mapped between that time and the late 1800s. Until the late 1930s, these maps were created by plane table. After that time, they were created from aerial photographs. No field notes were produced in the plane table surveys because angles were measured graphically and distances determined optically by stadia. Therefore, all work is shown on the manuscripts themselves.

On such maps, the high water line is the most prominent feature because it is used as the dividing line between land and water on the subsequently produced charts. The earliest specific instructions regarding the nature of the shoreline to be located were issued in 1898 (U.S. Coast Survey 1898); they called for the location of the "average high-water line" as opposed to the "stormwater line." The intent of such lines was apparently always that of the mean high water line although earlier mapping procedures may not have been as sophisticated as those used today.

These maps also depict the mean lower low water line and various other features, both natural and man-made, observed by the survey team. Section 4.4 provides additional information on the interpretation of these maps.

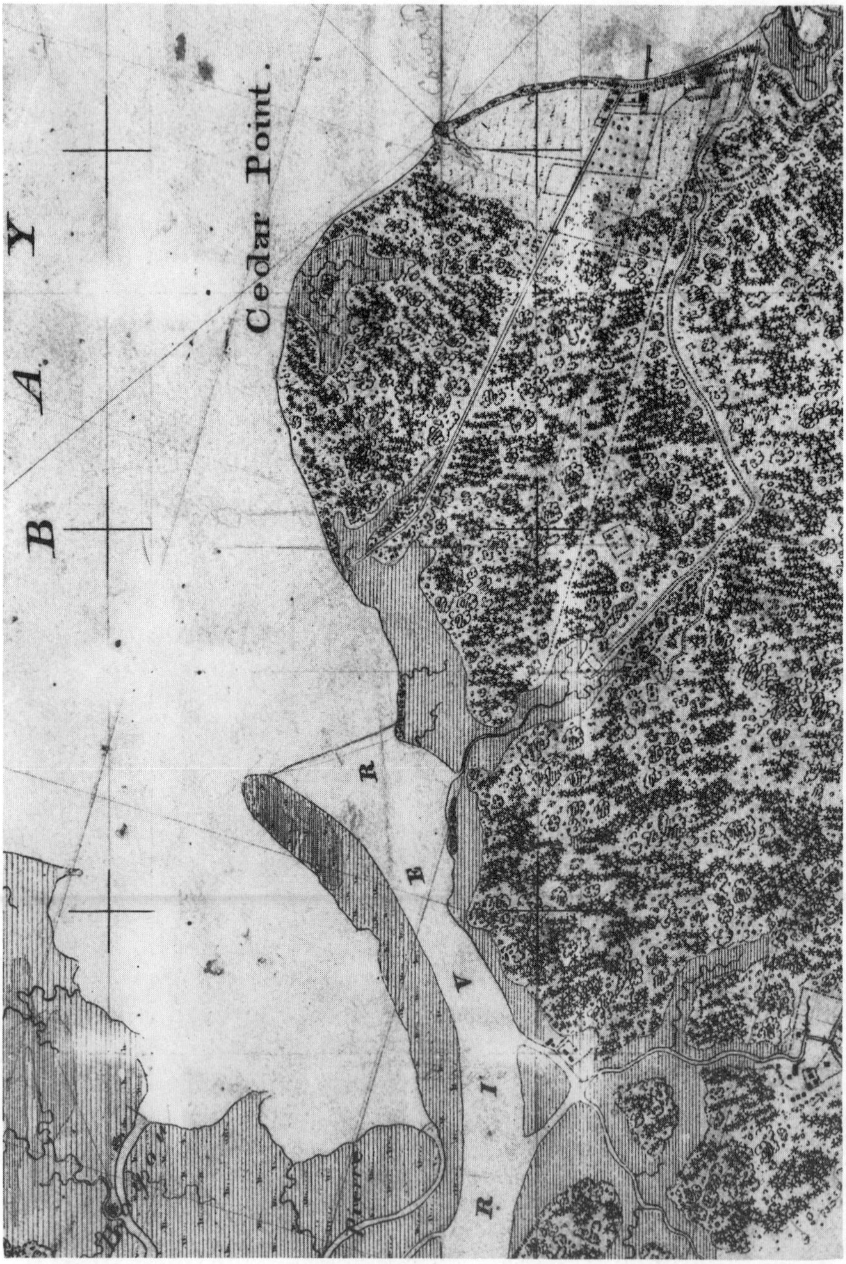

Figure 4.3 Typical "T" sheet (T370—Bay St. Louis, Mississippi, 1852).

4.3.2 Other Coast Survey Products

In addition to shoreline topographic maps, the U.S. Coast Survey, and its successor agencies, has produced a number of other products that may be of use in determining historic shoreline locations. These include hydrographic survey sheets, descriptive reports, and published nautical charts. Hydrographic survey sheets are field manuscripts containing plots of the various hydrographic soundings taken for the purpose of nautical chart production. Shorelines depicted on hydrographic sheets were usually traced from the accompanying topographic maps.

Descriptive reports were prepared by the surveyor directing the field surveys for the purpose of aiding the cartographer in preparing the final nautical chart. Such reports contain information helpful in interpreting and evaluating the accuracy of the topographic and hydrographic surveys. Often, information on the sources of the tidal data used in the project is included along with other information describing special conditions encountered. Nautical charts are the published maps resulting from the field survey data depicted on the topographic and hydrographic sheets. Usually, they are produced at a smaller scale than the field manuscripts. However, information sometimes may be found on the published charts, which is from sources other than the field sheets. Therefore, the charts can be a source of valuable information.

4.3.3 Bureau of Land Management Surveys

The Bureau of Land Management of the U.S. Department of the Interior and its predecessor agency, the General Land Office (GLO), has traditionally had the responsibility for surveying the public lands of the United States. Because much of the land within the current United States passed through federal ownership, surveys by these agencies represent the earliest systematic surveys in many areas of the country. Most of these surveys were for the purpose of subdividing federal territory into 6-mile-square townships and then into 1-mile-square sections for subsequent conveyance to private ownership. The product of such surveys are township plats and the field notes from which the plats were prepared.

When navigable bodies of water were encountered, the surveyors were instructed to run a meander line, or survey of the margin, of the water body. The purpose of such a meander line was to segregate the land that would remain in the public domain from land that was to be conveyed to private ownership. Meander lines were not intended to be precise mean high water line or ordinary high water line surveys. The meander points themselves may have been fairly accurately located on the high water line; however, they were infrequently located. Meander lines were intended to provide only a general representation of the shoreline. With an understanding of their limitations, however, these surveys are a useful source of historic shoreline location.

To aid in the interpretation and weighing of these surveys, it is often helpful to examine the instructions issued by the GLO to the deputy surveyors performing such surveys. The earliest of such instructions, issued by Surveyor General Tiffin in 1815, merely states that "for meandering rivers you will take the bearings according to the true meridian of the river and note the distance on any course when the river intersects the sectional lines. . . ." Subsequent instructions were generally more specific and can be useful in interpreting the survey.

4.3.4 Aerial Photography

There are a number of sources of systematic aerial photography. These include the National Ocean Service and its predecessor agencies, the U.S. Geological Survey, the Soil Conservation Service of the U.S. Department of Agriculture, and the National Aeronautics and Space Administration. In addition, aerial photography often may be obtained from various state agencies with repetitive photographic programs (such as transportation and revenue departments) and from private photogrammetric firms.

4.3.5 Data from Miscellaneous Sources

In addition to the specific sources listed before, historic maps are often available from numerous other sources. Some of these are the National Archives (information from numerous federal sources), the U.S. Corps of Engineers (hydrographic surveys, permit application files, and navigation studies), the U.S. Geological Survey (published maps, field notes, and hydrological studies), various state land offices, county court records, private surveying firms, title companies, and local historic societies.

4.4 INTERPRETATION OF HISTORIC SHORELINE MAPS AND SURVEYS

Because they are often the best available evidence of an historic position of a shoreline, some discussion regarding the interpretation of Coast Survey shoreline topographic maps is warranted. The following provides information on two frequently encountered problems in the use of these maps: interpretation of the shoreline and the handling of changes in horizontal datum.

4.4.1 Shorelines

As mentioned in previous sections, the high water line is the most important feature on such maps. Furthermore, this line represents the mean high water line as found by the topographer. One exception to this is in areas of tidal marshes. In such areas, the line of mean high water was not mapped. Instead, the seaward edge of the marsh was located (Shalowitz 1962; Swanson 1982). The reason for this is that the location of the line in such marsh areas could have been a difficult task and was not warranted for charting purposes.

Even though it was not intended to represent the mean high water line, the same symbol was used to represent the outer edge of marsh line as well as the mean high water line on most early surveys (Shalowitz 1962). Beginning in 1938, however, this line was indicated with a fine line as opposed to the standard weight line used to indicate the mean high water line. On modern shoreline manuscripts, the outer edge of marsh is usually labeled as the "apparent shoreline."

When marshes or grassy flats were encountered, which were mostly flooded at high water, this condition was often indicated on the map with a marsh symbol without a bordering line. In addition, the inner or upland edge of marsh was shown by a line on many of the earlier surveys, although this practice was discontinued on later surveys. Such a feature was not intended to represent the mean high water line and should not be interpreted as such.

In contrast to methods such as leveling, which might be used for precise boundary surveys, the high water line on shoreline maps was usually identified in the field by use of markings left on the beach by the preceding high water (Shalowitz 1962). With a knowledge of the tides in the area, this method was considered adequate for charting purposes. The line, ground located in this manner, was mapped by use of a plane table. This device is essentially a portable, tripod-mounted drafting board together with an alidade, which consisted of a telescope mounted on a ruler. The board was set up over a control point, the map manuscript secured to the board, and the board oriented so that the map corresponds to features on the ground.

Points along the shoreline were then located by sighting through the alidade at a graduated rod held on each shoreline point. By counting the number of graduations subtended by the stadia cross hairs in the telescope eyepiece, the distance to the shoreline point was determined. This distance then could be plotted along the line defined by the alidade ruler to map the shoreline point. Regarding accuracy of the position of these lines, it has been estimated that this method located the points on the shoreline with a maximum error of 10 meters (Shalowitz 1962). This estimate included error due to identification of the mean high water line on the ground, and many lines were located with considerable more accuracy than this. On modern day photogrammetricly derived shoreline sheets, the quoted accuracy is 5 meters.

4.4.2 Changes in Horizontal Datum

When attempting to compare shorelines mapped in early shoreline surveys with current day features, it is necessary to use a common horizontal datum. All such maps have grid ticks for latitude and longitude. However, there have been several

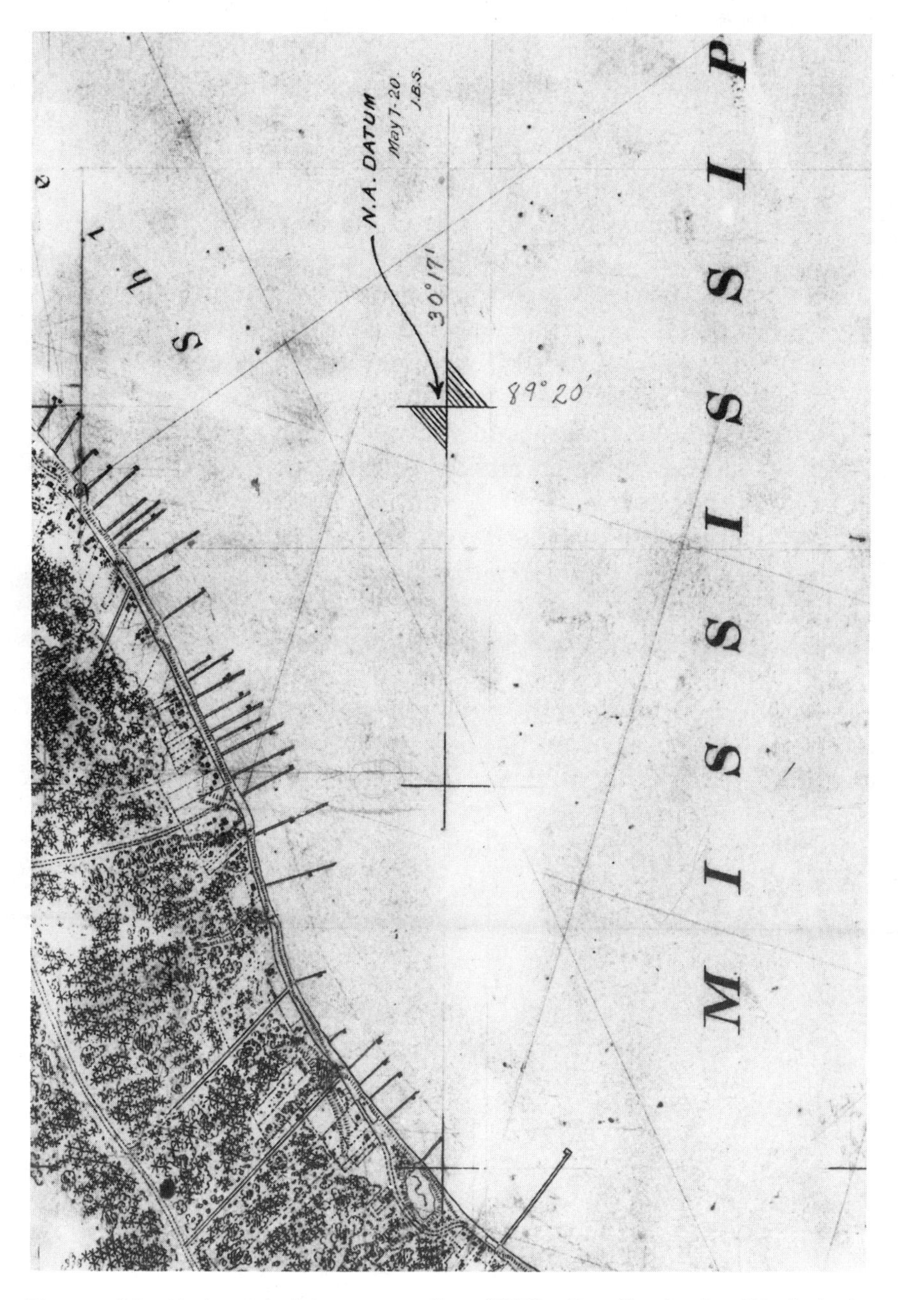

Figure 4.4 Horizontal datum correction (T370—Bay St. Louis, Mississippi, 1852).

changes in the reference spheroid, in methods of field measurement of latitude and longitude, and in the national horizontal datum. Therefore, to compare a map compiled at an earlier time with a current map, the relationship between the two systems used must be known.

Generally, hand corrections have been made to early topographic surveys to allow for changes in geographic position. This was usually done by adding corrective grid ticks adjacent to the original ticks. Figure 4.4 shows a typical correction. The multiple grid ticks represent successive changes due to different reference system. For reliable comparison with information from other topographic maps or with current features, care should be taken to ensure that the same reference system is used for all data. For example, currently, most Coast Survey topographic maps have corrective ticks for the North American Datum of 1927. However, data for National Geodetic Survey control stations are currently published on the North American Datum of 1983. Therefore, either 1927 values for the control stations should be used in mapping current features when comparing them with historic shorelines or a correction to the 1983 datum made to the projection ticks on the historic topographic map. Data are available from the National Geodetic Survey (National Ocean Service, NOAA) for determining differences between various datum systems.

5

USE OF GOVERNMENT LAND OFFICE MEANDER LINES AS BOUNDARIES*

5.1 INTRODUCTION

One of the more important charges to government surveyors performing the original surveys of public lands of the United States was to separate the upland, subject to disposal under the public land laws, from the navigable waters, which were to be preserved as public highways. Even a century or two later, the decisions of these surveyors take on a great deal of significance in disputes regarding lands lying under and bordering on water bodies. The separation process involved mean-

* This section was originally written as an article in *Surveying and Land Information Systems,* Vol. 50, No. 3, September 1990, American Congress on Surveying and Mapping (Cole 1990) and is reprinted with permission.

dering, or running a series of measured courses and distances around the perimeter of the water body. The alternate course of action, for nonnavigable water bodies, was to continue section lines across the water bodies. For such nonnavigable water bodies, the width was usually noted in the field notes and the approximate limits of the water sketched on the plat, as opposed to an actual survey or meander of the shoreline, as required for navigable water bodies.

5.2 TO MEANDER OR NOT TO MEANDER

5.2.1 Pertinent Instructions

The meandering of navigable water bodies has been an integral part of the survey of the public lands since its inception (Bouman 1977). As an illustration of this, the first survey conducted under the rectangular survey system (Township 1, Range 1, Seven Ranges, Ohio by Deputy Surveyor Absolom Martin) included meandering of the Ohio River. However, the early laws detailing procedures for public land surveys did not address meandering as such. They did, however, mention the reservation of navigable rivers. This is illustrated by the following:

> That all navigable rivers, within the territory to be disposed of by virtue of this act, shall be deemed to be, and remain public highways. . . . *(Section 9, Act of May 18, 1796)*

The earliest detailed instructions for the public land survey were those issued by Surveyor General (Northwest of the Ohio) Edward Tiffin in 1815. These instructions did provide some direction as to which waters to meander, at least for rivers, but did not address lakes. The instructions read as follows:

> The courses of all navigable rivers, which may bound or pass through your district, must be accurately surveyed. . . .

The first mention of lakes appears to be in the instructions for the survey of lands in the state of Mississippi, issued in

1832 by Gideon Fitz, surveyor general of the lands south of Tennessee. In 1842, instructions to deputy surveyors in Florida also mentioned lakes:

> You will accurately meander, by course and distance, all navigable rivers which may bound or pass through your district; all navigable bayous flowing from or into such rivers; all lakes and deep ponds of sufficient magnitude. . . .

The first general nationwide manual of instructions, issued in 1855, provided even more specific language regarding rivers, and included a requirement to meander all lakes of 25 acres or larger. A published supplement to the 1855 manual, written in 1864, modified the manual to increase the lower limit for lakes to 40 acres. In addition, a new class of nonnavigable river was mentioned. The instructions related to those two items read:

> Rivers not embraced in the category denominated "navigable" under the statute, but which are well-defined natural arteries of internal communication and have a uniform width, will be meandered on one bank. . . .
>
> Lakes embracing an area of less than forty acres will not be meandered. Long, narrow or irregular lakes of larger extent, but which embrace less than one-half of the smallest legal subdivision, will not be meandered. Shallow lakes or bayous, likely in time to dry up or be greatly reduced by evaporation, drainage, or other cause, will not be meandered, however, extensive they may be.

The requirement for meandering of one bank for nonnavigable "arteries of internal communication" was also included in the 1881 manual, and then dropped in later manuals. The 1881 manual also did away with the requirements to meander all lakes of 40 acres and above, and instead required that "all lakes, bayous, and deep ponds which may serve as public highways of commerce" be meandered. Those instructions also required that "lakes, bayous, and ponds lying entirely within a section are not to be meandered." The 1890 manual of instructions reinstated the requirement that all lakes over

25 acres were to be meandered. This restriction was also included in the 1894, 1902, 1919, and 1930 manuals. All size restrictions were deleted in the last two (1947 and 1973) manuals.

The 1890 manual also instituted the so called three-chain rule for streams. This instruction required the meandering of streams wider than three chains (198 feet) wide, regardless of their navigability. This rule was also included in the 1894, 1919, and 1930 manuals, but was deleted in later manuals.

5.2.2 Legal Significance of Meandering

As may be seen from the previous example, federal surveyors made an effort to identify all navigable waters in public lands. These surveys were often performed relatively near the time of statehood, so they reflect navigability determinations of the water bodies in the condition they were in at that time. Therefore, this would be the deciding factor in determining which water bodies were received by the states by virtue of statehood. Because of this, these surveys often represent the most reliable and often the only inventory of navigable waters in most public land states. Therefore, the courts have relied on the public land surveys for a presumption of navigability or nonnavigability. An example of this is a recent Florida Supreme Court case *(Odom v. Deltona):*

> In Florida, meandering is evidence of navigability which creates a rebuttable presumption thereof. The logical converse of this proposition, noted by the lower court, is that nonmeandered lakes and ponds are rebuttably presumed nonnavigable.

Obviously, the early surveyors were not infallible—any more than surveyors are today. Therefore, their decisions regarding the navigability or nonnavigability of a water body were not always correct. As indicated in the previous opinion, such evidence creates only a presumption, which may be rebutted by evidence to the contrary. Nevertheless, these surveys are obviously the best evidence of the navigability of a water body at statehood in most cases, and therefore, they have received judicial recognition as such.

However, when considering the evidence of meandering in government surveys, one must be aware of the prevailing instructions controlling that survey. For example, the occasional requirement that lakes down to a size of 25 acres be meandered, regardless of their navigability, indicates that a meandered lake would not be necessarily considered as navigable.

5.3 LOCATION OF MEANDER LINES

5.3.1 Pertinent Instructions

Although, as has been demonstrated in previous sections, meandering was mentioned in some of the earliest laws and instructions, directions as to placement of the meander line were not mentioned until the 1881 manual. Salient portions of that manual follow:

> Meander lines should not be established at segregation line between dry and swamp or flowed land, but at the ordinary low-water of the actual margin of the rivers or lake border. . . .
>
> In the survey of lands bordering on tide water, "meander corners" are to be established at points where surveyed lines intersect high-water mark, and meanders are to follow the high-water line.

These instructions remained the same in the 1890 manual. In the 1894 manual, the instructions remained the same, except that the term "ordinary low-water mark" (in the first paragraph cited before) was changed to "ordinary high-water mark." The 1919 manual reflected considerably more detail in the meandering section and contained instructions that have remained essentially the same in subsequent manuals. Salient portions of those instructions follow:

> All navigable bodies of waters and other important rivers and lakes (as hereinafter described) are to be segregated from the public lands at mean high water elevation.
>
> Mean high-water mark has been defined in a State decision (47 Iowa, 370) in substance as follows: High-water mark in the Mississippi River is to be determined from the river bed; and that

only is riverbed which the river occupies long enough to wrest it from vegetation.

In another case (14 Penn. St., 59) a bank is defined as the continuous margin where vegetation ceases, and the shore is the sandy space between it and low water mark.

Meander lines will not be established at the segregation line between upland and swamp or overflowed land, but at the ordinary high water mark of the actual margin of the river or lake on which such swamp or overflowed lands border.

Practically all inland bodies of water pass through an annual cycle of changes from mean low-water to flood stages, between the extremes of which will be found mean high-water.

The surveyor will find the most reliable evidence of mean high-water elevation in the evidence made by the water's action at its various stages, which will generally be found well marked in the soil, and in timbered locations a very certain indication of the locus of the various important water levels will be found in the belting of the native forest species.

Mean high-water elevation will be found at the margin of the area occupied by the water for the greater portion of each average year; at this level a definite escarpment in the soil will generally be traceable, at the top of which is the true position for the surveyor to run the meander line.

5.3.2 Legal Significance of Location

From the previous example, the earliest instructions regarding the location of meander lines (1881) indicate that these lines were to be established at what we today consider to be the ordinary high water mark. It may be assumed that those instructions represented the codification of previous practices. Therefore, this probably had been the practice since the inception of the public land surveys.

The detailed instructions in the later manuals are for a line that is identical to ordinary high water lines located by today's practicing surveyors in nontidal waters. In tidal waters, the instructions obviously do not yield as precise and repeatable a line as a more sophisticated mean high water line determined by modern techniques. Nevertheless, in most cases, the line found by such instructions would be a good approximation of the mean high water line. Therefore, it appears that the meander line was at least intended to be, and should be in most cases, a reasonable approximation of the ordinary high water line or mean high water line at the time of the survey. Water boundaries are often ambulatory, so this obviously affects the use of meander lines as current boundaries. This is reflected in a number of court opinions typified by the following:

> It is an established and accepted principle, subject only to the exception hereinafter noted, that the meander line of an official government survey does not constitute a boundary, rather the body of water whose shoreline is meandered is the true boundary. . . . (Connerly v. Perdido Key, Inc.)

Meander lines are significant, however, in that they may represent the approximate location of the ordinary high water line at a certain point in history. This is often a factor in determining the original location of altered shorelines. In addition, there may be other instances when the current water boundary cannot be located and the meander line may be accepted as the boundary by default, or where the discrepancy between the meander line and the shoreline is large enough to indicate intentional omission of certain lands or fraud. These instructions are illustrated by the following excerpts from court opinions:

> Under the circumstances of this case (testimony had indicated that the true mean high-water line could not be located with any certainty), we hold that the meander line constituted the boundary line between the swamp and overflowed lands and the sovereignty lands. . . . (Trustees v. Wetstone)

> However, a meander line may constitute a boundary where so intended or where the discrepancies between the meander line and the ordinary high-water line leave an excess of unsurveyed land so great as to clearly and palpably indicate fraud or mistake. *(Lopez v. Smith)*

Therefore, although certain limitations must be recognized, the location of a meander line indeed can have legal significance.

6

WHICH WATERS ARE SOVEREIGN?

6.1 GENERAL CRITERIA

An obvious question arising when defining the boundary between sovereign waters and private uplands is "Which waters are sovereign?" A simplistic answer to the question is navigable waters. However, that is not an explicit answer because there are many definitions of navigability. In addition, there are some water bodies that are navigable-in-law although not necessarily navigable-in-fact, and others are navigable-in-fact although not necessarily sovereign.

There are three general criteria for a water body to be considered sovereign. The first is that it must be a natural water body. The second is that it must have been considered navigable and in federal ownership at the time of statehood of the state in which it is located, and therefore conveyed to that state by virtue of statehood. The third is that it must not have been subsequently conveyed by the state to other parties.

6.2 NAVIGABILITY-IN-FACT VERSUS NAVIGABILITY-IN-LAW

In nontidal waters, navigability for title purposes generally is a question of navigability-in-fact. In Florida, for example, recent case law *(Odom v. Deltona)* offers specific clarification to that state's definition in nontidal waters. In that case, the court held that "Florida's test for navigability is similar, if not identical, to the Federal Title Test." The Federal Title Test was defined as being "based on the body's potential for commercial use in its ordinary and natural condition."

Further understanding of the Federal Title Test may be obtained from the following early statement of this test *(The Daniel Ball):*

> And they are navigable in fact when they are used or susceptible of being used in their ordinary condition, as highways for commerce, over which trade and travel are or may be conducted in the customary modes of trade and travel over water.

It appears from the preceding and subsequent cases that the Federal Title Test does not allow consideration of artificial improvements to the water body. It must be navigable in its ordinary and natural condition. The test has been clarified to indicate that waters are considered navigable even if the waters are not navigable during the entire year. *(Bucki v. Cone; U.S. v. 2,899.17 Acres of Land Etc.).* However, the implication has been that the lack of navigability is an exceptional circumstance as opposed to the predominant condition.

In tidal waters, navigability for title purposes is not always based on navigability-in-fact. In some states, public ownership extends to submerged lands subject to the ebb and flow of the tide, regardless of actual navigability. An excellent in-depth coverage of this subject (Maloney and Ausness 1974) indicates that in Louisiana, Maryland, Mississippi, New Jersey, New York, and Texas, state ownership extends to all waters subject to tidal ebb and flow, whereas in California, Connecticut, Florida, North Carolina, and Washington, public ownership is based on navigability-in-fact. The same article states that "Alabama, Oregon and South Carolina find tidal watercourses prima facie navigable and thus presume the land beneath the watercourses to be sovereign land, but this presumption of state ownership may be rebutted by a finding of nonnavigability."

As noted in the preceding list of navigability-in-fact states, Florida is generally considered to be such a state. There is at least one Florida case that reflects a different opinion. *Martin v. Busch* states that "the State, by virtue of its sovereignty, became the owner of all lands between the ordinary low water mark and ordinary high water mark, and also all the *tidelands,* viz. lands covered and uncovered by the daily ebb and flow of normal tides" (emphasis added). However, the preponderance of Florida case law appears to support the opinion reflected in Maloney and Ausness (1974). For example, *Clement v. Watson* states that "waters are not under our law regarded as navigable merely because they are affected by the tide." Another case *(City of Tarpon Springs v. Smith)* states this opinion in the same words. A more recent case *(Board of Trustees v. Wakulla Silver Springs Co.)* also reflects this opinion (Tucker 1983). Furthermore, the navigability-in-fact position is clearly implied in Section 177.28 of the Florida Statutes, which states that the ". . . mean high water line along the shores of land *immediately bordering* on navigable waters is recognized and declared to be the boundary between the foreshore owned by the state in its sovereign capacity and upland subject to private ownership" (emphasis added).

As mentioned before, Mississippi has been generally considered to be an ebb-and-flow state. However, there has been a recent challenge *(Cinque Bambini Partnership et al. v. State of Mississippi et al.)* as to whether all tidally affected waters are sovereign regardless of navigability-in-fact. The author served as the state's expert in that case. At the chancery court level and in the Mississippi Supreme Court, this case was decided in favor of ebb and flow. Salient excerpts from the State Supreme court opinion are as follows:

> The early federal cases refer to the trust as including all lands within the ebb and flow of the tide.
>
> . . . it is our view that as a matter of federal law, the United States granted to this State in 1817 all lands subject to the ebb

> and flow of the tide and up to the mean high water level, without regard to navigability.
>
> Yet so long as by unbroken water course—when the level of the waters is at mean high water mark—one may hoist a sail upon a toothpick and without interruption navigate from the navigable channel/area to land, always afloat, the waters traversed and the lands beneath them are within the inland boundaries we consider the United States set for the properties granted the state in trust.

The case was appealed to the U.S. Supreme Court *(Phillips Petroleum Co. v. Mississippi),* which concurred with the state court in ruling that all coastal states received all lands over which tidal waters flow, and that Mississippi still does. The opinion noted that this ruling "will not upset tides in all coastal states" because it "does nothing to change ownership rights in states which previously relinquished a public trust claim to tidelands such as those at issue here." Therefore, it is presumed that the ruling has no net effect in those states that have established navigability-in-fact case law. However that issue has yet to be tested.

6.3 NONNAVIGABLE COVES AND TRIBUTARIES

In many situations, there are nonnavigable waters, such as coves and tributary streams, adjacent to navigable water bodies. A question arises as to whether such waters are part of the navigable water body or are separate water bodies requiring their own determination of navigability. This section addresses that issue.

The primary application of such a determination is in sovereign/upland boundaries. Obviously, this is not an issue in tidally affected waters for those jurisdictions recognizing tidal influence as the criteria for navigability-in-law. In navigability-in-fact jurisdictions, however, the answer to this question determines whether such adjoining waters were merely portions of the navigable water body, and therefore in sovereign ownership, or separate nonnavigable water bodies, and thereby subject to private ownership.

In a typical situation, this issue involves a nonnavigable indentation, such as a cove, adjoining a larger, navigable water body. All of the submerged lands associated with the larger water body are sovereign (unless conveyed by the state), including areas near the shore where the lack of depth precludes navigation. The mean high water line (or the ordinary high water line in nontidal water bodies) along the shores of the larger water body is the limit of these sovereign lands. The question at issue is whether the nonnavigable area within the indentation is considered as being within the reaches of the larger water body and thus sovereign land.

A logical approach to such a dilemma is to apply the same criteria to this situation as in those determinations for defining bays covered in Chapter 7. If the indentation meets those tests, then it should be considered a separate water body. It therefore is presumably nonnavigable and subject to private ownership. The indentation would be severed from the larger water body by a straight line from headland to headland. The mean high water line is the boundary between sovereign and private lands only in the larger water body and not within the indentation. Conversely, if the indentation does not meet the prescribed tests, the submerged lands within would be considered part of the larger water body and therefore sovereign. In that situation, the mean high water line is the sovereign/private boundary within the indentation as well as in the main part of the larger water body.

If the indentation is a nonnavigable stream, as opposed to a nonnavigable bay, it would seem logical to apply the techniques suggested in Chapter 7 for determining the closing line across the mouth of rivers. The limits of the sovereign submerged lands of the larger water body are then determined by the straight line drawn between the headlands of the stream. If the stream is navigable, sovereign ownership would extend up the stream as far as it is navigable-in-fact.

In some areas, indentations are found with a pronounced berm or bar across their entrances. In such a situation, the key issue is the water connection, if any, between the larger water body and the navigable-in-fact water within the indentation. If the berm is intact with no navigable connection at mean high water, then the submerged land within the indentation is not considered sovereign. If the berm contains nonnavigable breaks, then the same situation exists. However, if a natural, navigable passage exists through the berm, then the submerged lands within the indentation could be considered as sovereign subject to the limitation expressed in the previous three paragraphs. The inland extent of sovereign ownership is then the mean high water line inside the indentation.

In Florida, this approach has been substantiated by the State Supreme Court. In one case *(Clement v. Watson),* the court found that submerged land inside a cove was not sovereign. The entrance to the cove was blocked by a berm through which the only navigable passage was a man-made channel. In another case *(Baker v. State ex rel Jones),* the Florida Supreme Court held that a nonnavigable arm of a navigable lake was not sovereign in character and thus was not subject to public use.

Where the entrance to the indentation is obstructed by dense, impenetrable vegetation, rather than a solid berm or bar, there seem to be no clear guidelines. However, one authority (Maloney and Ausness 1974) states that "if the vegetation is really impenetrable, it might well be equated with a berm that prevents navigation, in effect making the cove a separate nonnavigable water body and perhaps, therefore, subject to private ownership."

6.4 FLOODPLAINS ADJOINING NAVIGABLE WATERS

In many riverine systems, especially in regions with low relief, there are wide, low-lying floodplains paralleling the navigable channel of the river. Usually in such systems, there is a natural levee between the channel of the river and the floodplain. In some areas, the floodplain contains heavily vegetated swamps inundated during the flood stages of the river. Some areas of the floodplain are lower than the levee. Water from flood stages of the river or upland runoff, or water that has entered through breaks in the levee, is often entrapped in such areas for considerable portions of the year.

At times, questions have arisen regarding whether low-lying

portions of the floodplain, or even the entire floodplain, are considered sovereign land. In some states, such areas have been the subject of administrative claims of sovereign ownership. The rationale for such claims is that such areas may be lower than the ordinary high water line; and that although evidence of the ordinary high water line may be found at the bank of the river, additional evidence of a similar nature may be found at the upland edge of the floodplain.

Despite such claims, case law appears to clearly require the exclusion of the floodplain from the sovereign owned bed of the water body. Examples of this are as follows:

> It neither takes in overflowed land beyond the bank, nor includes swamps or low grounds liable to be overflowed but reclaimable for meadows or agriculture. . . . *(Borough of Ford City v. United States; Howard v. Ingersoll)*

> . . . nor the line reached by the water at flood stages. *(State ex. rel. O'Conner v. Sorenson)*

> It is the land upon which the waters have visibly asserted their dominion, and does not extend to or include that upon which grasses, shrubs and trees grow, though covered by the great annual rises. *(Harrison v. Fite)*

> The high water mark on fresh water rivers is not the highest point to which the stream rises in times of freshets. . . . *(Dow v. Electric Company; Tilden v. Smith)*

Further clarification as to the floodplain question may be obtained from the 1849 and 1850 Acts of Congress that granted swamp and overflowed lands to many of the states. Swamp lands are defined as lands that "require drainage to fit them for cultivation. Overflowed lands are those which are subject to such periodical or frequent overflows as to require levees or embankments to keep out the water, and render them suitable for cultivation" *(San Francisco Savings Union v. Irwin)*. "Overflowed lands include essentially the lower level within a stream flood plain as distinguished from the higher levels . . ." (Bureau of Land Management 1973). Based on these definitions, the floodplain appears to fall into the classification of swamp and overflowed land rather than sovereign land.

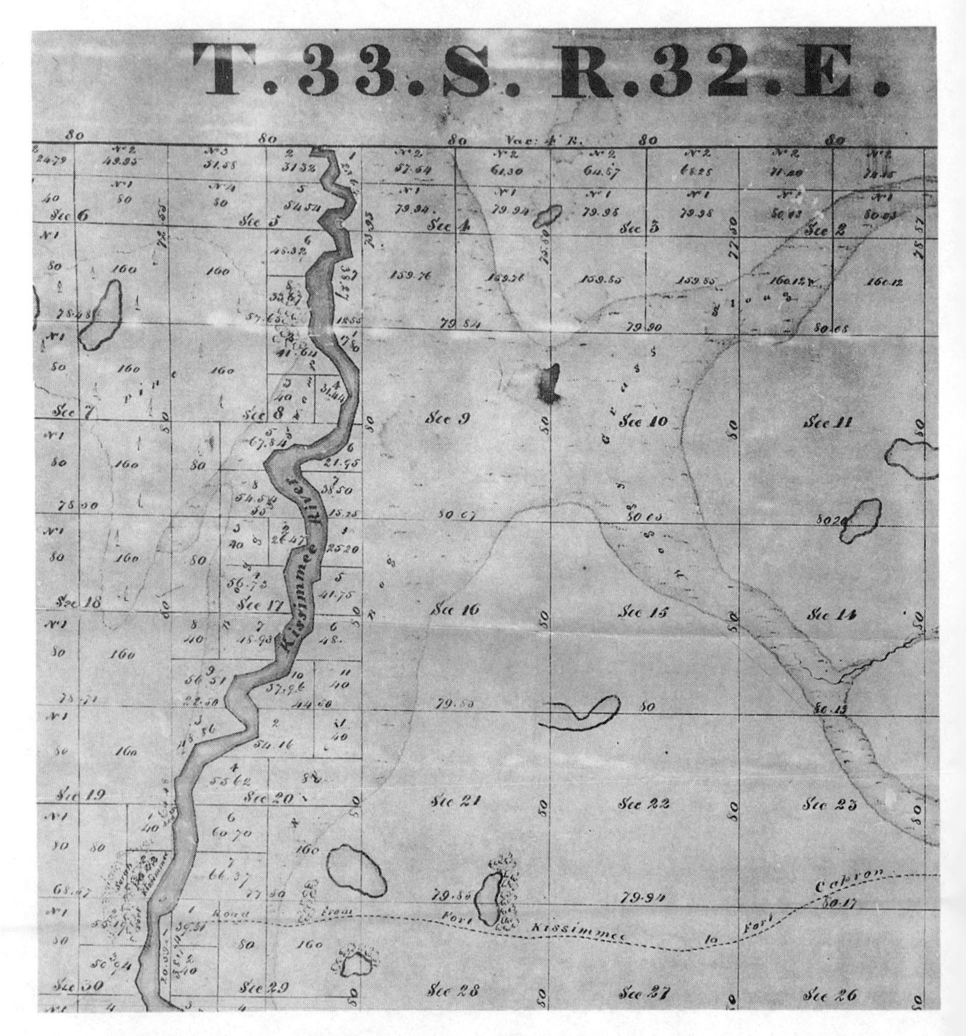

Figure 6.1 A portion of an 1845 Government Land Office (GLO) plat.

Examination of the original General Land Office (GLO) surveys further substantiates the exclusion of the floodplain from sovereign land. In most cases, the original surveyors meandered along the limits of the bed of the water body and not the edge of the floodplain. Figure 6.1, a copy of an 1845 GLO plat, shows how the floodplain was typically excluded from the navigable river. Earlier guidelines to the GLO surveyors, as well as current instructions, direct such a location. For example, the 1868 manual of government surveying (Hawes 1868) addresses this issue as follows:

> . . . care must be taken in time of high water, not to mistake the margins of bayous or the borders of overflowed marshes or "bottoms" for the true river bank.

7

BOUNDARIES BETWEEN WATER BODIES

7.1 INTRODUCTION

To the uninitiated, the question of where one water body ends and another begins seems trivial—somewhat akin to deliberations over how many angels can occupy the head of a pin. To those involved in water boundary disputes, however, this is often a difficult and intensely contested issue. The question as to the limits of water bodies can have a significant impact on international boundaries as well as state/federal and sovereign/upland boundaries.

The discussion of this issue is divided into three areas. The first includes dividing lines between adjacent navigable waters; the second covers limits of navigable water bodies with adjacent nonnavigable waters; and the third provides information on the unique problems of finding the extent of sovereign ownership in ebb-and-flow jurisdictions where the limit of tidal influence is the inland boundary.

7.2 BOUNDARIES BETWEEN ADJACENT NAVIGABLE WATER BODIES

7.2.1 Applications

In a number of legal situations, it is necessary to precisely delineate the boundary between adjacent water bodies. This is especially true along coastlines with indentations such as bays and rivers. Often such indentations are subject to regulations different from those for the main water body. These regulations include fishery statutes, water quality standards, boating ordinances, and others. Such indentations also may be subject to separate ownership as when the indentation or tributary is nonnavigable.

As discussed in later chapters, closing lines across coastal indentations are also used as a baseline from which to measure specific distances for delineating federal/state as well as international maritime boundaries. Therefore, the position of the closing lines across these coastal indentations obviously can be of importance in the enforcement of various regulations or in determining limits of jurisdiction or resource interest.

The procedures to be presented for establishing these closing lines were developed primarily for international boundaries. However, they represent informed scientific and legal thought as to the limits of water bodies. Because of this, they form a strong precedent for application in national, state, and local jurisdictions, and in a number of cases have been specifically applied in judicial decisions other than in international law. Therefore, they are considered valid for determining the limits of most water bodies.

7.2.2 Definition of Bays

Whether a coastal indentation is a separate water body or merely a part of the open sea creates a unique question in dealing with federal, state, and international offshore boundaries. If the indentation is indeed a bay and therefore part of the inland waters, then the baseline for delineating the offshore boundary is a straight line across the entrance. If the indentation is part of the open sea, then the baseline follows the shoreline. Thus, the criteria used for defining bays can be critical. The Convention on the Territorial Sea and the Contiguous Zone (CTSCZ)—from United Nations Law of the Sea Conference in 1958—defines a bay as a "well marked indentation where penetration is in such proportion to the width of its mouth as to contain landlocked waters and constitute more than a mere curvature of the coast" (Paragraph 2, Article 7, CTSCZ). Therefore, that a water boundary is called a bay does not necessarily make it a juridical bay or a bay in the legal sense. From this definition, the extent of the penetration of the waters into the land, in proportion to the width of the entrance, appears to be the major criterion.

To quantify this criterion in objective mathematical terms, the so called semicircle rule was developed and added to the preceding definition: "An indentation shall not, however, be regarded as a bay unless its area is as large as, or larger that, that of the semicircle whose diameter is a line drawn across the mouth of that indentation." For this purpose, the entrance width is measured between the low water line of its natural entrance points and the area lying below the low water line around the shore of the bay. Islands within the area are included as if a part of the water area. Figure 7.1 illustrates the semicircle rule. Figure 7.1(a) depicts an indentation that does not meet this rule whereas 7.1(b) depicts one that does.

Another approach for comparing the ratio of penetration of the coastal indentation to the width of its mouth was suggested by the International Law Commission in its input to the CTSCZ and is helpful as background for such determinations. This approach suggested that ". . . the distance between the two extremities shall be twice the depth of the indention." This, of course, is the same ratio as a semicircle. Determining the depth is obviously somewhat subjective because there are a number of ways in which this could be accomplished. To objectively determine this depth, one source (Hodgson and Alexander 1972) recommends use of the longest straight line

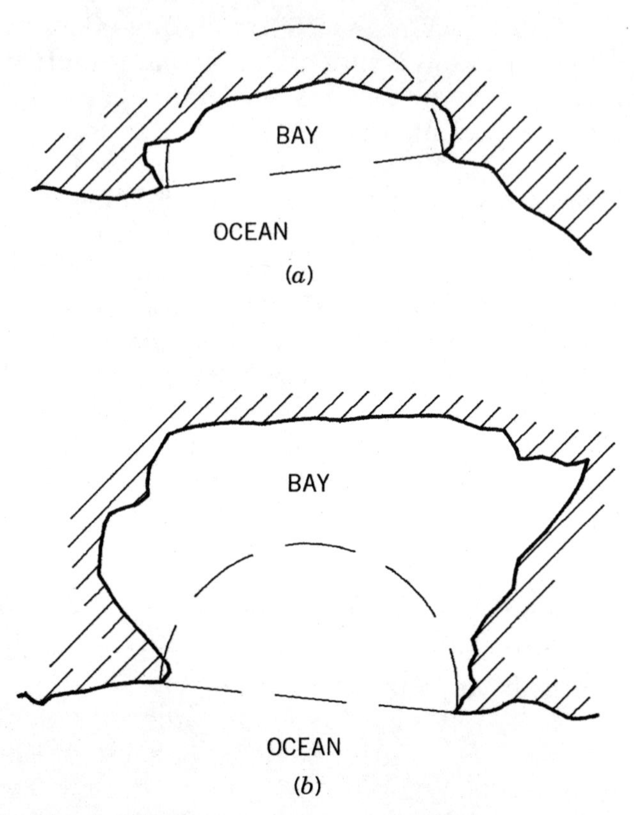

Figure 7.1 (a) Indentation not meeting the semicircle test. (b) Indentation meeting the semicircle test (Hodgson and Alexander 1972).

that can be drawn from any point on the closing line to the head of the indentation. Figure 7.2 depicts this technique along with other techniques for determining depth of penetration.

When islands are located in the entrance, the situation is somewhat more complex. The Convention requires that "the semicircle shall be drawn on a line as long as the sum total of the lengths of the lines across the different mouths." Figure 7.3 shows a typical configuration with entrance islands.

A further limitation was placed on the definition of a bay by the provision that the entrance to a bay does not exceed 24 nautical miles in width (Paragraph 4, Article 7, CTSCZ). There-

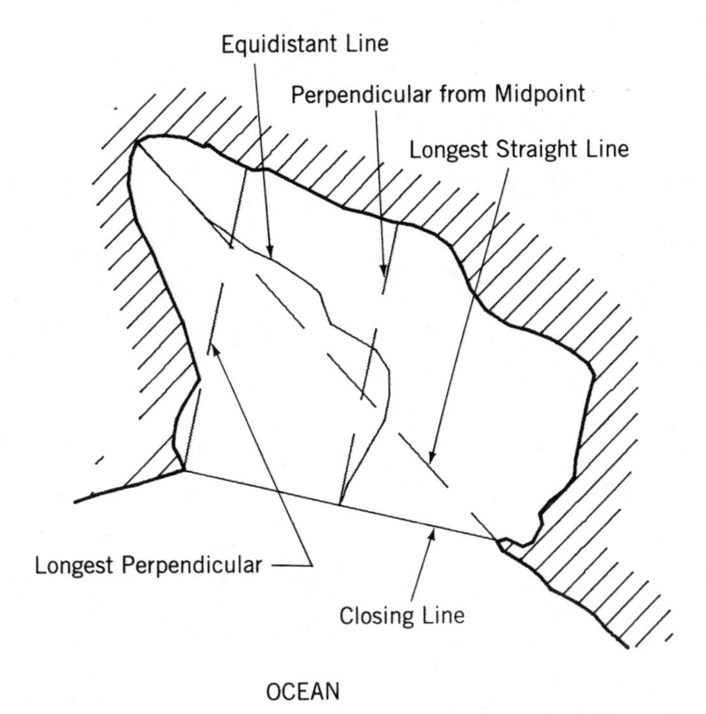

Figure 7.2 Determining penetration distance (Hodgson and Alexander 1972).

fore, indentations with wider entrances are not considered as bays unless a portion of the indentation meets the criteria of a bay in itself. In that case, the smaller portion is defined as a bay. Figure 7.4 shows this restriction.

There is one class of indentations that is regarded as an exception to the rule defining juridical bays. These indentations are historic bays, areas for which, through long-standing assertion of rights and acquiescence by others, a nation has established a prescriptive title. Such claims may have been established over large bodies of water not meeting the normal criteria for bays. The obvious purpose of this exception is to exclude from examination ". . . certain bays whose status has been already settled by history" (Shalowitz 1962).

Where rivers enter the sea, a partition line is drawn straight across its mouth (Article 13, CTSCZ). Because no maximum width is specified, this is apparently true regardless of the width of the entrance.

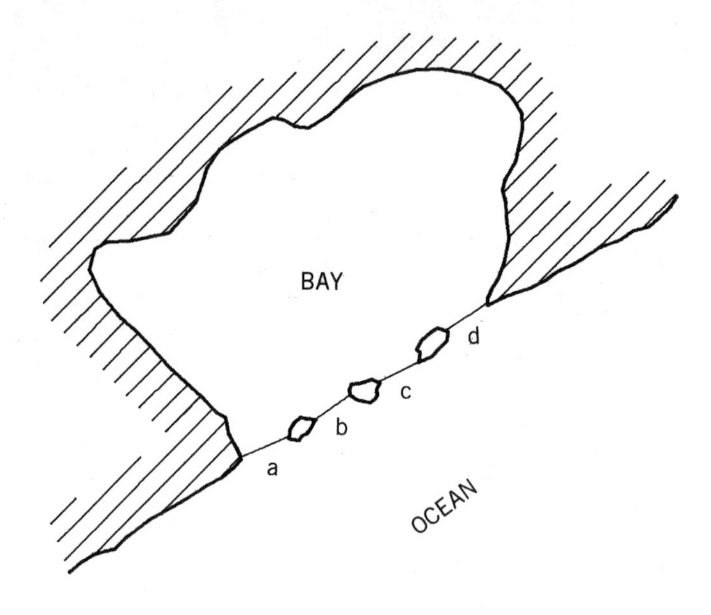

Figure 7.3 Typical configuration with entrance islands (Beazley 1978).

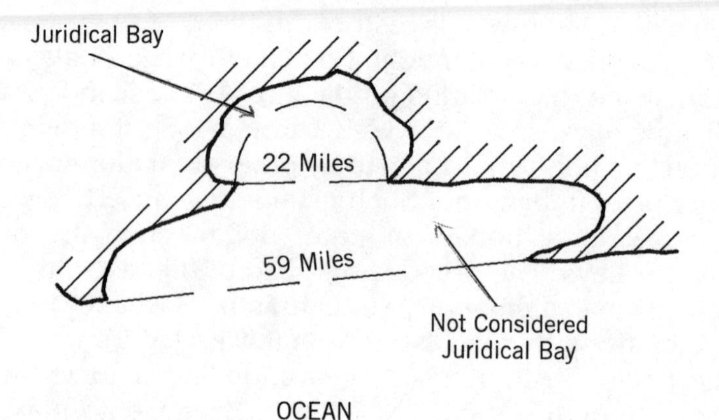

Figure 7.4 An entrance exceeding 24 miles with interior portions defined as a bay (Shalowitz 1962).

7.2.3 Entrance Points (Headlands)

For both bays and rivers, the seaward limit of such water bodies is a headland to headland line. The principal problem in the application of this, however, is the selection of acceptable entrance points or headlands. For water bodies with pronounced physical features as headlands, experts tend to closely agree. For other water bodies, however, there is often a wide disparity in opinion.

In common usage, the word headland implies a pronounced elevation. Webster defines a headland as "a point of usually high land jutting out into a body of water." In the context used for water boundaries, however, the horizontal dimensions are more important than elevation. In this context, a headland is defined as ". . . the apex of a salient of the coast; the point of maximum extension of a portion of the land into the water; or a point on the shore at which there is appreciable change in direction of the general trend of the coast" (Shalowitz 1962). Shalowitz further comments that the shoreline of headlands is formed by the meeting of the forces of the ocean and those of the estuary. The combination of these forces frequently form some characteristic point, such as a sand spit or cusp, that would be the headland sought. Such points should be the outermost extension of the headland.

When the headland does not have a distinct point and is more rounded, then a point should be selected by bisecting the angle formed by the lines following the general trend of the low water line of the open coast and the low water line of the estuary. Figure 7.5 shows the bisector of the angle approach. The U.S. Supreme Court *(U.S. v. California)* specifically endorsed this method as follows:

> Where there is no pronounced headland, the line should be drawn to a point where the line of mean lower low water on the shore is intersected by the bisector of the angle formed where a line projecting the general trend of the line of mean lower low water along the open coast meets a line projecting the general trend of the line of mean lower low water along the tributary estuary. *(U.S. v. California)*

Hodgson and Alexander (1972) suggested an objective method of selecting the correct entrance points of a indenta-

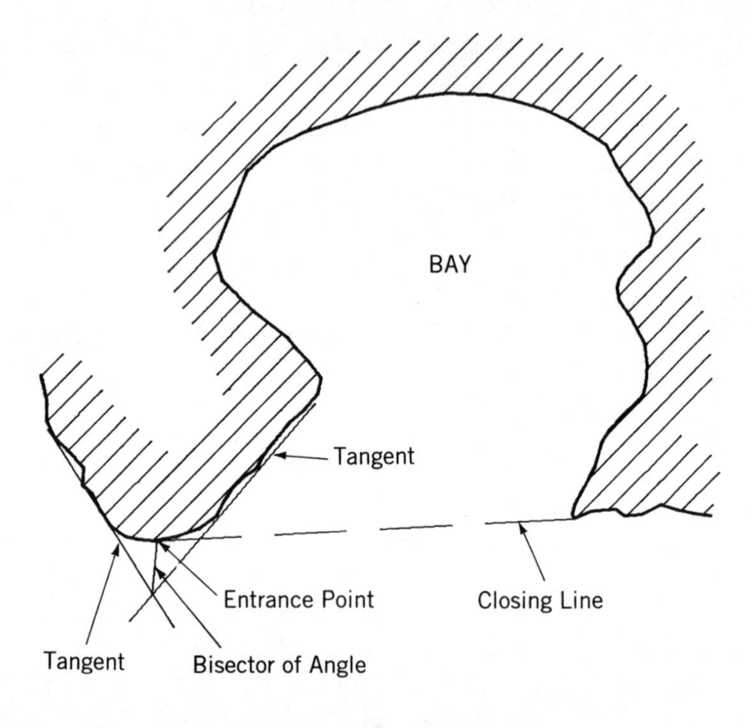

Figure 7.5 Bisector of angle approach for locating entrance point on rounded headland (Shalowitz 1962).

tion using a "45-degree test." Because 45 degrees represents a dividing line between perpendicular vectors, this test uses that criterion for locating the dividing line between the opposing lines of the ocean and the indentation. If the angle between the general direction of the shoreline within the indentation and the proposed closing line is 45 degrees or greater, then, by this test, the correct headland has been selected (see Figure 7.6). Although this test has not been specifically endorsed by the U.S. Supreme Court, it has been mentioned in dictum in one case *(U.S. v. Maine).* As with most criteria, one must use this test with care. As pointed out in Hodgson and Alexander (1972), there are certain configurations of bay entrances for which the indiscriminate use of this test would result in a point not at the true outermost limit of an indentation. Figure 7.7 shows one such exception. Point *A* does not meet

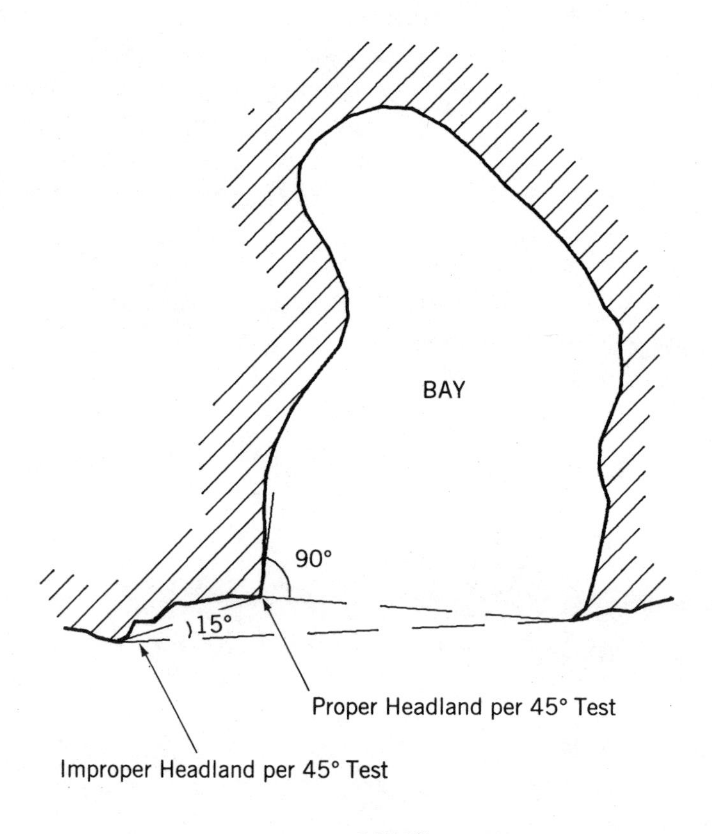

Figure 7.6 Forty-five-degree test for headland selection (Hodgson and Alexander 1972).

the 45-degree test, whereas Point *B* does. Yet, Point *A* is the correct headland.

7.2.4 Obstructed Entrances

When islands or other land masses occur near the entrance to indentations, this may result in complications in the selection of the entrance points. The correct treatment for such islands is the subject of some disagreement because this situation is not addressed specifically in the CTSCZ. However, reasonable interpretations are made consistent with the intent of the CTSCZ as expressed in other situations.

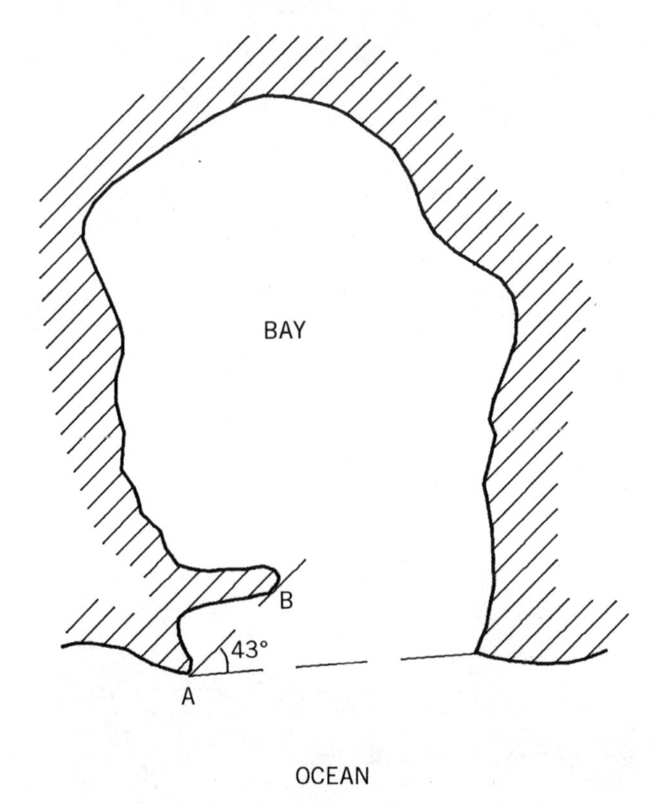

Figure 7.7 Erroneous results using the 45° test (Hodgson and Alexander 1972).

The first step in a decision as to how to treat entrance islands is to draw a tentative closing line disregarding the islands. If the islands are intersected or are outside the line, as shown in Figures 7.8(a) and 7.8(b), then the closing line should be redrawn using the headlands of the islands. This is consistent with Article 7, Paragraph 5 of the CTSCZ, which indicates that the maximum area of water possible should be enclosed. Shalowitz (1962) concurs with this approach but points out that the question of how far seaward the islands can be in order to be incorporated under this approach is unanswered. Hodgson and Alexander (1972) concur with this approach for islands intersected by the original closing line and for "screening islands" located outside the original closing

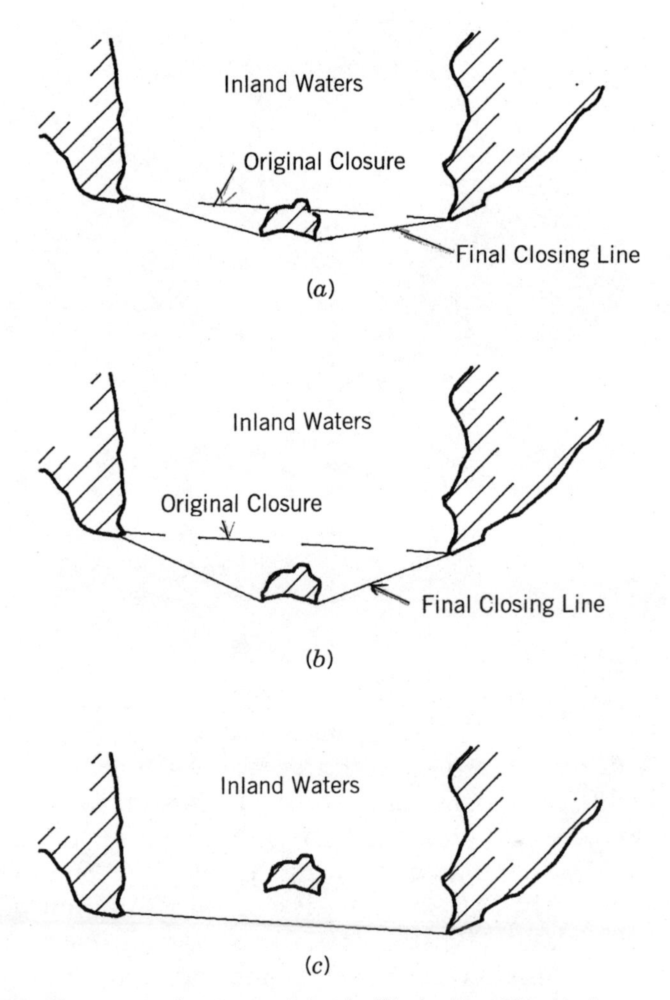

Figure 7.8 Treatment of entrance islands (Shalowitz 1962; Hodgson and Alexander 1972).

line. (Screening islands are defined as those islands that block more than one-half of the entrance.) However, Hodgson and Alexander make no mention of other entrance islands outside the original closing line.

For islands lying within the original closing line, see Figure 7.8(c), the correct treatment disregards the islands. This is consistent with Paragraph 3, Article 7 of the CTSCZ, which

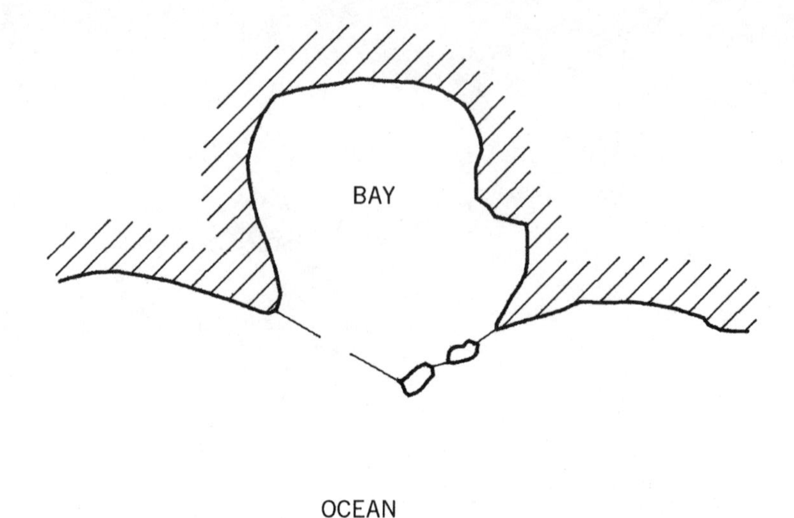

Figure 7.9 Use of entrance islands as headlands (Beazley 1978).

allows islands within a coastal indention to be considered as part of the water area. Hodgson and Alexander (1972) apparently are in disagreement with this approach and report that even if inside the closing line, screening islands (as previously defined as constituting more than 50 percent of the entrance width) must be used as part of the closing line. Therefore, by that approach, the closing line is concave, dipping down to connect the islands. Beazley (1978) specifically disagrees with Hodgson and Alexander on that point and states that ". . . in general, if the single closing line satisfies the semicircle rule there is no compelling reason to treat the islands as anything but islands lying within a bay."

When entrance islands are closely related to the adjoining mainland, they themselves may constitute the headlands of the indentation. This is the case when islands act as a natural prolongation of the mainland, such as shown in Figure 7.9.

When a peninsula from within the indentation exists near the entrance, creating a multiheaded bay, the logical treatment is to enclose both of the adjacent bays with a single closing line. Figure 7.10 illustrates this situation.

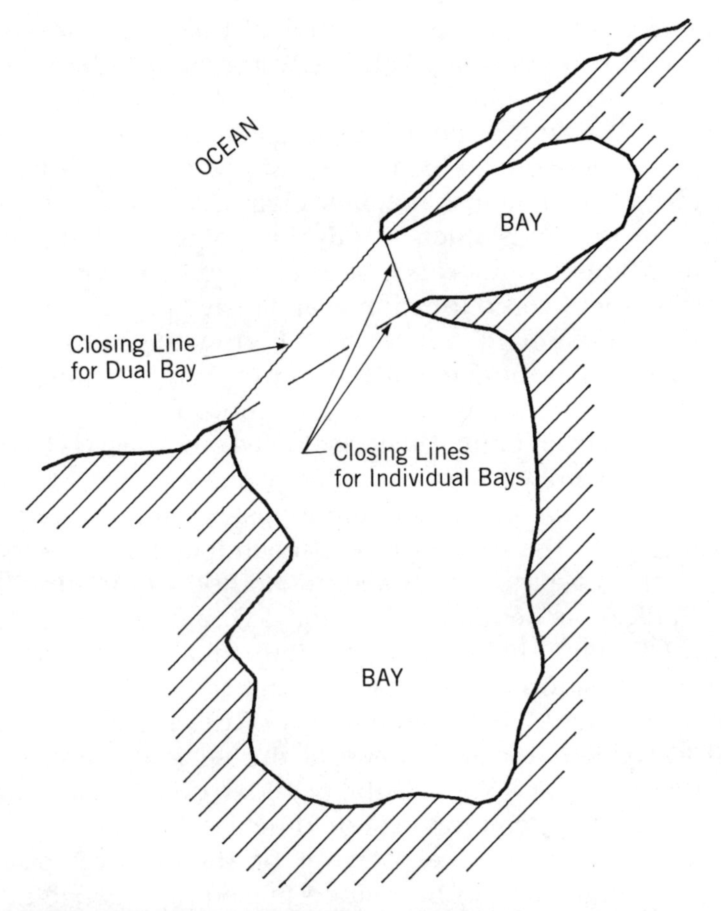

Figure 7.10 Two-headed bay (Hodgson and Alexander 1972).

7.3 LIMITS OF TIDAL INFLUENCE

In jurisdictions where all waters subject to the ebb and flow of mean high tides are considered navigable-in-law regardless of navigability-in-fact, a unique problem exists in finding the inland limit of sovereign ownership. In those ebb-and-flow jurisdictions that claim only tidally affected waters, the inland extent of sovereign ownership in all water courses is the limit of tidal influence, sometimes called the *head of tide.* In those jurisdictions that claim all tidally affected waters plus navigable-in-fact nontidal waters, the limit of tidal influence represents the inland boundary in those water courses that are not navigable-in-fact.

Locating the limit of tidal influence can be a complex problem. The objective is to locate the point where tidal influence ends on a tide that just reaches mean high water. Unfortunately, very few tides reach exactly that point and then recede. Most tides either do not reach or else exceed mean high water. Therefore, the desired location is an illusive point that can be very time-consuming to locate. Fortunately, many tidal streams are tidally affected all the way to their headlands, thus eliminating the necessity for this location.

To locate such a point, the most logical approach is to begin by setting several gauges at random points upstream. By observing a few tidal cycles simultaneously with an established control station, the mean range at each gauge may be calculated by use of either the Standard Method or the Amplitude Ratio Method, as detailed in Section 1.3.5. of this text. Then, if the relationship between the various mean ranges appears linear, the position of the inland extent of tidal influence can be approximated. Misleading values of the mean range may result for gauges located inland of the desired point. Therefore, to eliminate this possibility, tidal cycles with ranges greater than the mean range should be used with care.

Once the approximate location of the desired point is found, the position can be refined by setting a gauge at that point together with other gauges located short distances upstream and downstream from that point. Once again, tidal observations should be made simultaneously with an established control station and mean ranges calculated. By repeating this process a number of times, the inland extent of tidal influence may be found with precision, depending on the number of reiterations.

8

STATE AND FEDERAL WATER BOUNDARIES

8.1 BACKGROUND AND HISTORY

In earlier sections, it was stated that U.S. case law has upheld state ownership of inland waters. Through the early part of the twentieth century, it was generally believed that this same rule applied to the marginal sea (Hanna 1951). This was evidenced even by official actions of Department of Interior in the mid-1930s (Briscoe 1984). However, in the 1930s, with the discovery of oil in the submerged lands off the California coast, disputes arose as to the ownership of such lands. The resulting conflict led to a series of cases called the Tidelands Decisions *(U.S. v. California, U.S. v. Louisiana,* and *U.S. v. Texas).* These cases held that the federal government, rather than the states, was owner of the margin sea.

Following the Tidelands Decisions, considerable pressure was reportedly exerted on the U.S. Congress by various coastal states. This culminated in the passage of the Submerged Lands Act *(43 U.S.C., s 1301-1 1970)* in 1953. This act relinquished federal interest in the marginal sea within each state's boundaries.

Therefore, each coastal state now holds title to a band of submerged land bordering its coastline. It is the seaward boundary of this land that we now address.

8.2 BOUNDARY DEFINITIONS

The earliest concept of the closed or marginal sea evolved as a matter of self-defense on the part of various coastal nations. The territorial sea of a nation was, for practical purposes, the zone that the nation could comfortably defend for its exclusive use from hostile shipping. In the eighteenth century, this was more or less standardized by the so-called "cannon shot" rule. By that informal rule, the boundary of the marginal sea was the distance from shore that a cannon shot could reach. This was generally considered to be 3 miles. This same historic definition has become the basic rule, for the most part, for the current seaward boundary of state ownership.

The Submerged Lands Act defines this boundary in the following two sections:

> Section 2(b). The term "boundaries" include the seaward boundaries of a State or its boundaries in the Gulf of Mexico or any of the Great Lakes as they existed at the time such State became a member of the Union, or as heretofore approved by the Congress, or as extended or confirmed pursuant to Section 4 hereof but in no event shall the term "boundaries" or the term "lands beneath navigable waters" be interpreted as extending from the coast line more than three geographical miles into the Atlantic Ocean or the Pacific, or more than three marine leagues into the Gulf of Mexico.
>
> Section 4. The seaward boundary of each original coastal State is hereby approved and confirmed as a line three geographical miles distant from its coastline or, in the case of the Great Lakes, to the international boundary. Any State admitted subsequent to the formation of the Union which has not already done so may extend its seaward boundaries to a line three geographical miles distant from its coastline or to the International boundaries of the United States in the Great Lakes or any other

> body of water traversed by such boundaries. Any claim heretofore or hereafter asserted either by constitutional provision, statute, or otherwise, indicating the intent of a State claim so as to extend its boundaries is hereby approved and confirmed, without prejudice to its claim, if any it has, that its boundaries extend beyond that line. Nothing in this section is to be construed as questioning or in any manner prejudicing the existence of any State's seaward boundary beyond three geographical miles if it was so provided by its constitution or laws prior to or at the time such State became a member of the Union, or if it has been heretofore approved by Congress.

Two states on the Gulf of Mexico are especially affected by the last sentence in the preceding excerpt. The issue of whether any of the Gulf Coast states were entitled, under the Act, to submerged lands greater than three geographical miles from the coastline was decided by the Supreme Court in 1960 *(U.S. v. Louisiana, Texas, Mississippi, Alabama and Florida).* That decision held that Texas and Florida were entitled to submerged lands extending 3 leagues into the Gulf due to the extent of their boundaries at the time of admission into the Union. However, it was held that Louisiana, Mississippi, and Alabama were entitled to a marginal sea of only three geographical miles.

More recently, the Supreme Court has continued to clarify more complex issues regarding state/federal boundaries at various locations around the country. Included among these are *U.S. v. Louisiana,* which in 1969 applied the principles of the 1954 Geneva Convention to these boundaries; *U.S. v. Alaska,* which in 1975 found Cook Inlet to be neither a juridical nor historic bay (therefore the state/federal boundary is measured from the shoreline inside the inlet rather than from a closing line); *U.S. v. California* (1977), which established closing lines across entrances to several bodies of inland waters; *U.S. v. Florida,* which determined the status of Florida Bay as inland water and which defined where the Atlantic Ocean ended and the Gulf of Mexico (with its 3-league marginal sea) began; and *U.S. v. Louisiana* (Mississippi Sound case), which in 1985 determined the ownership of "enclaves" or pockets within Mississippi Sound located more than 3 miles from both the mainland shoreline as well as from the shore-

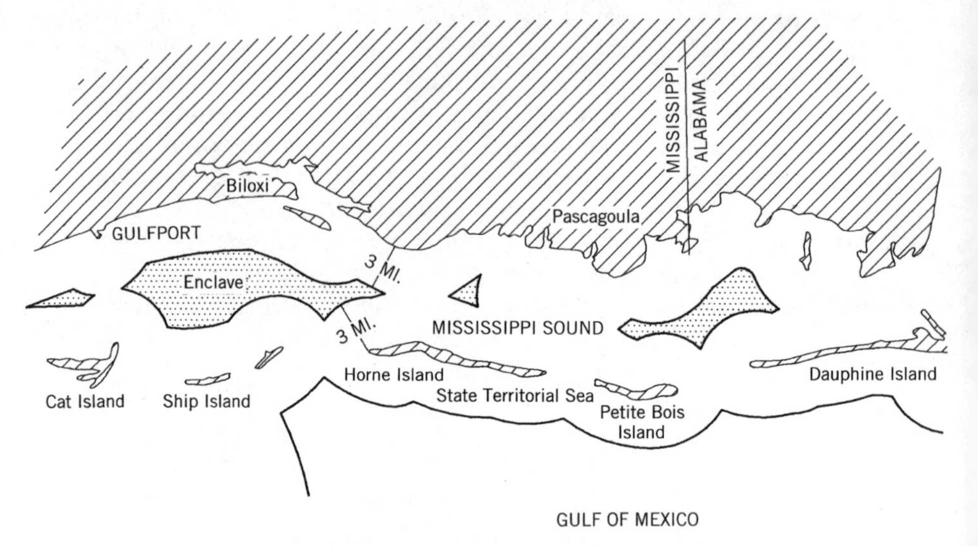

Figure 8.1 Mississippi Sound enclave dispute configuration.

line of barrier islands bordering the sound. Figure 8.1 shows this configuration. The Supreme Court found that Mississippi Sound was both a juridical bay as well as an historic bay. Therefore, the waters of Mississippi Sound were inland waters and entirely owned by the states.

In *U.S. v. Florida,* it is interesting to note that the Court found the dividing line between the Atlantic Ocean and the Gulf of Mexico to be a line running due west from the Marquesas Keys along latitude 24° 35′ north. Therefore, the seaward boundary for Florida is three geographical miles from the coastline in the Atlantic and on the south side of the Florida Keys to a point at latitude 24° 35′ north. At that point, the boundary goes due west to a point three marine leagues from the coastline. Then the boundary follows the coastline, at 3 leagues distance, northerly and westerly along the Gulf coast. A separate envelope, 3 miles wide to the south and 3 leagues wide to the north exists around the Dry Tortugas Keys. Figure 8.2 shows the boundary in this area.

For clarification, several terms used in this discussion need definition:

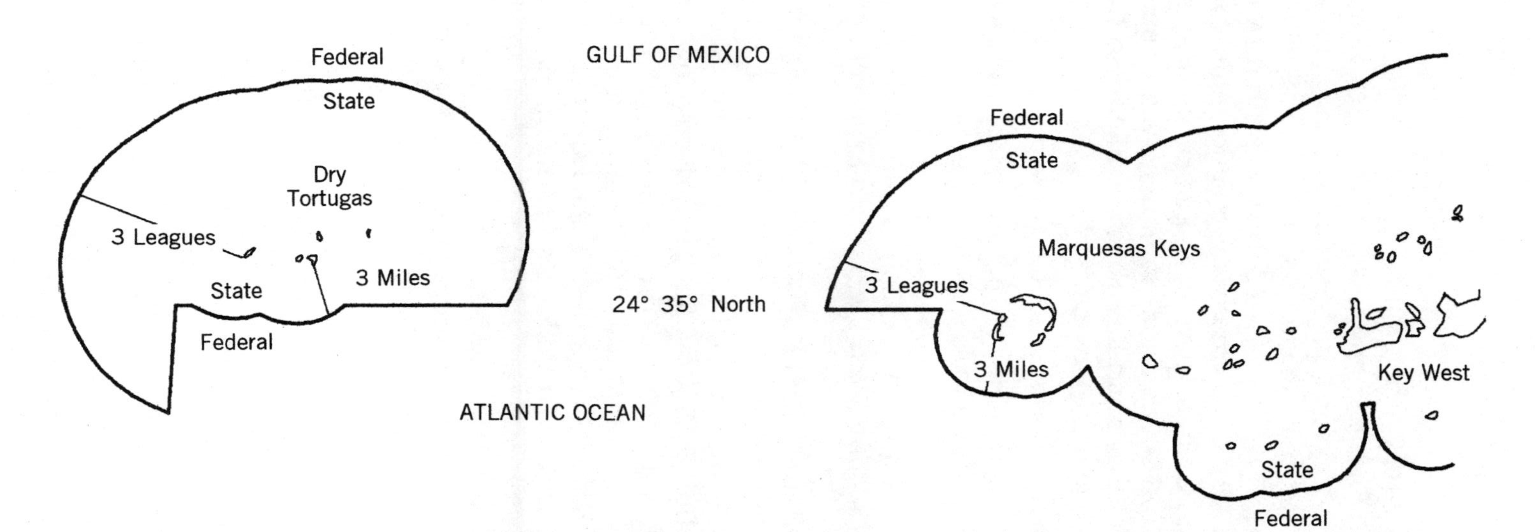

Figure 8.2 State/Federal boundary around Dry Tortugas Keys.

Geographical Mile: A geographical mile is the length of 1 minute of arc on the equator or 6087.09 feet on the Clarke Spheroid of 1866 (Bowditch 1962). It is noted that this is slightly different than an international nautical mile which is approximately 6076.11549 U.S. feet. It is interesting that in the 1947 *U.S. v. California* case, a geographical mile was defined by the Court in dictum as being a nautical mile. Further, the federal government commonly uses nautical miles for measuring this boundary which is slightly to the disadvantage of the states involved.

Marine League: A marine league is three geographical miles.

Coastline: "The coastline for the purpose of measuring out to the seaward boundary is . . . the line of ordinary low water along that portion of the coast which is in direct contact with the open sea and the line marking the seaward limit of inland water" *(Section 2 (c), Submerged Lands Act)*

The "seaward limit of inland water" in the definition refers to any closing line drawn across bays, river mouths, or other inland waters. A more complete discussion of such closing lines is in Section 7.2. The "line of ordinary low water," or baseline for measuring out to the boundary, is interpreted as the low water line along the coast, as marked on large-scale charts of the coastal state (Article 3, First Geneva Conference on the Law of the Sea). Traditionally for the United States, this is the mean lower low water on the Pacific Coast and the mean low water on the Atlantic Coast. Recently, however, this has been changed to mean lower low water on both coasts (National Ocean Survey 1980).

8.3 TECHNIQUES FOR LOCATING STATE/FEDERAL WATER BOUNDARIES

For most purposes, the seaward water boundary of state submerged land may be delineated, as suggested in the previous section, by drawing arcs 3 miles (or 3 leagues where appropriate) from salient points on the lower low water line on large-scale nautical charts. That line, as drawn, then can be physically located by use of navigational systems.

Whenever it is desirable to know the relationship of a point or area in relation to such a boundary with greater precision than allowed by the aforementioned method, a four-step process should be used:

1. Establish a precise datum for lower low water along the coastline in the area of concern.
2. Map the line of mean lower low water in this area in relation to a common coordinate system (often, the line need be located only in areas around salient points).
3. Select, either graphically or numerically, a series of salient points on the mean lower low water line that control the location of a line three geographical miles (or 3 leagues where appropriate) seaward.
4. Compute the coordinates for the boundary line itself and, as needed, measure coordinates for the point or area in question.

Steps 1 and 2 basically follow techniques described in Section 1.3. A local datum for mean lower low water is first established. Then, the intersection of that elevation with the rising shore is located and mapped. That line then serves as the baseline from which the location of the offshore boundary is computed. Where there are rivers or bays, closing lines across the entrances may be drawn, as is the manner described in Section 7.2. Likewise, where there are offshore shoals, islands, or rocks that uncover at mean lower low water, the seaward most part of the mean lower low water line of those would control as salient points from which to measure to the actual boundary.

9

LATERAL WATER BOUNDARIES OF STATES

9.1 TYPES OF LATERAL BOUNDARIES

Lateral boundaries, as used herein, refer to those projecting seaward from the coastline between adjacent waters or states. Delimitation of such boundaries involves the same problems for both international and interstate seaward lateral boundaries. For both, such delimitation has as its objective the division of the submerged land in a manner equitable to both parties to the boundary. Developing rules to achieve this in all cases can be difficult. The simplest method for drawing lateral boundaries may be used where the shoreline is fairly straight and the land boundary between the states is perpendicular to the shoreline. For such a configuration, the most logical and equitable division line is a seaward extension of the land boundary. However, such an ideal configuration rarely exists, and the straight-line extension approach could result in inequitable solutions under some configurations. Figure 9.1 shows

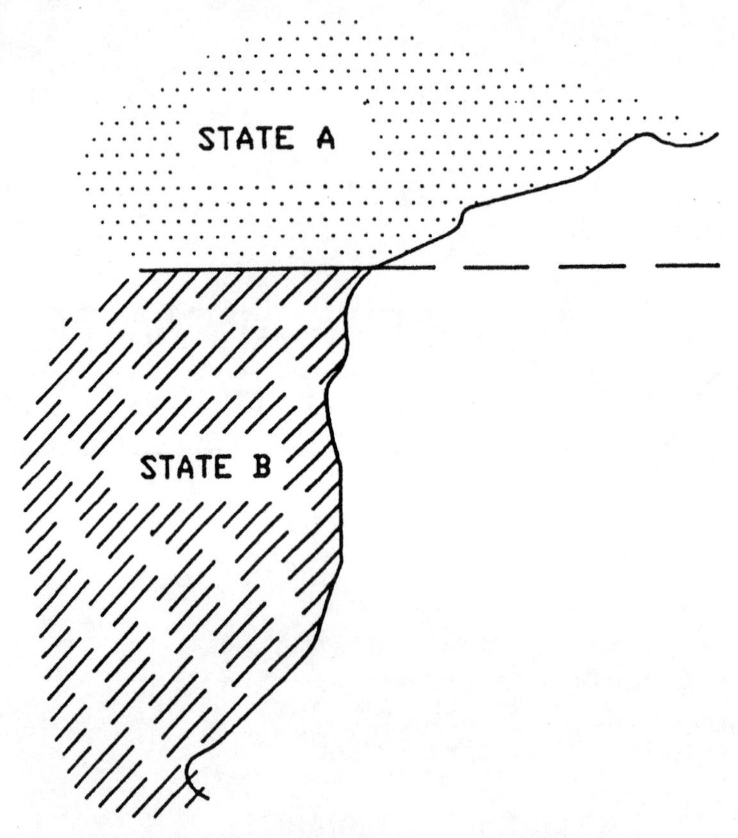

Figure 9.1 Inequitable solution from the straight-line extension of land boundary.

such a situation. From that figure, a straight-line extension results in denying one state the right to submerged land to which it is obviously entitled. For the configuration in the figure, an obviously more equitable division line is one more perpendicular to the shoreline at the termination of the land boundary.

To accommodate various shoreline/land boundary configurations, various other approaches have been developed. These include (1) a line projected at right angles to the coast where the land boundary reaches the shore; (2) a line projected at right angles to the general trend of the coast; (3) a line projected in a cardinal direction; and (4) an equidistant line. A detailed explanation of the construction of an equidistant line is found in Section 9.2.

For international lateral boundaries, an equidistant line is specifically mandated by the Convention on the Territorial Sea and Contiguous Zone adopted by the 1958 Geneva Convention on the Law of the Sea. Article 12 of the Convention states:

> Where the coasts of two states (nations) are opposite or adjacent to each other, neither of the two States is entitled, failing agreement between them to the contrary, to extend its territorial sea beyond the median line every point of which is equidistant from the nearest point on the baselines from which the breadth of the territorial sea of each of the two States is measured.

In the true sense of the word, an equidistant line boundary is not a median line. A median line is a line through the territorial sea between two coastal nations that lie opposite one another (Shalowitz 1962). However, such a line has the same underlying principle as the equidistant lateral boundary. Therefore, both types of boundaries are commonly called median lines. From the earlier excerpt, the Geneva Convention combines both cases (adjacent coasts and opposite coasts) into the same article.

For interstate lateral boundaries, the procedures required by the Geneva Convention are not binding. Due to this, no one method has been used exclusively. Rather, whichever of the previously mentioned options providing the most practical solution has been selected. Such boundaries may have been established by one of the state's original charter or grant, by interstate agreement between the states party to the boundary, or by judicial decree.

Except where an international boundary is involved, the federal government has no authority to determine state lateral boundaries in the territorial sea. Where state boundaries are extended beyond the territorial sea for special purposes, however, the federal government has claimed exclusive right to delimit such boundaries (Christie 1979). For example, the Outer Continental Shelf Lands Act (OCSLA) codified the Truman Proclamation (see Section 10.1) in extending U.S. and state jurisdiction to the seabed and subsoil of the continental shelf, and directed the President to "determine and publish in the *Federal Register* such projected lines extending seaward

. . ." *(43 U.S.C. § 111333(a)(2))*. Another example is the Coastal Energy Impact Program (CEIP), which is an amendment to the Coastal Zone Management Act. The CEIP provided for allocation of funds connected with oil and gas production on the outer continental shelf. That allocation is based on the amount of shelf area adjacent to each coastal state. To determine that area, the CEIP authorized the National Oceanic and Atmospheric Administration (NOAA) to project the lateral boundaries to the limit of the continental shelf, subject to subsequent modification by interstate agreement *(16 U.S.C. § 1456a(b)(3)(B))*.

9.2 CONSTRUCTION OF EQUIDISTANT AND MEDIAN LINES

For the purposes of this discussion, an equidistant line for either adjacent or opposite coasts is referred to as a median line. Figure 9.2 shows the construction of a median line boundary for adjacent coasts. For that configuration, the median line from the termination of the land boundary between States A and B to the limits of the territorial sea is constructed as follows (Shalowitz 1962):

1. An offshore point, outside of the territorial sea, is selected. This offshore point (Point 1) should be equidistant from salient shoreline points (identified on the figure as Points a and b) for both states.
2. From Point 1, the median line runs shoreward along a line that is a perpendicular bisector of line ab (between the two salient points).
3. The median line continues shoreward until Point 2 is found that is equidistant from Points a and b as well as from the next nearest salient point (Point c) on the coast of either state.
4. From Point 2, the median line runs shoreward along a line that is a perpendicular bisector of line cb.
5. As shown in Figure 9.2, this process is continued, with the median line running shoreward along the perpendicular bisectors from the various equidistant points en-

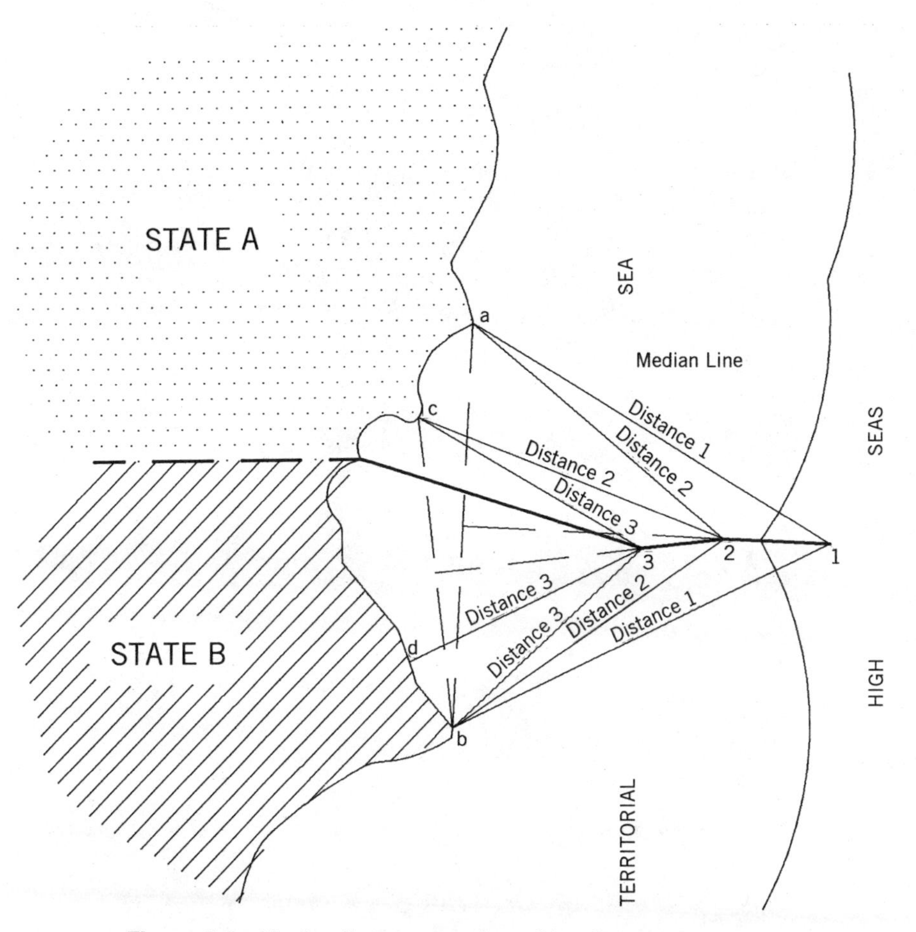

Figure 9.2 Median line construction with adjacent coasts.

countered, until the land boundary on the coast is encountered. The last course obviously may be other than a perpendicular bisector.

Median line construction may be complicated by the presence of offshore islands. Figure 9.3 shows a typical configuration of a median line for adjacent coasts with offshore islands (Pearcy 1959). Such islands complicate the determination but do not prevent a median line from being applicable, as seen from the illustration.

When a median line is to be established between coasts op-

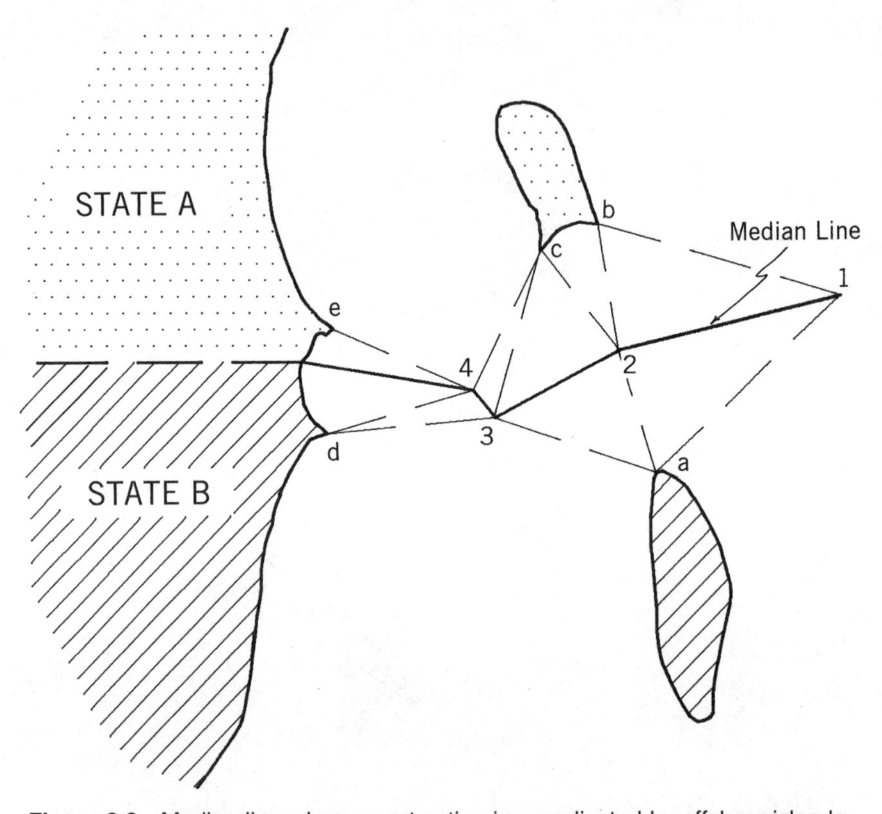

Figure 9.3 Median line where construction is complicated by offshore islands.

posite each other, a slightly different procedure is followed. Figure 9.4 shows the construction of such a median line. For that configuration, the median line between States A and B is constructed as follows (Shalowitz 1962):

1. An initial point should be established. To locate an initial point, first select a prominent point on each coast. These points (a and b) must be selected so that the chosen point on the coast of State A (Point a) is the closest point on that coast to Point b. Likewise, Point b must be the closest point on the coast of State B to Point a. The midpoint between Points a and b is the initial point (Point 1) of the median line.

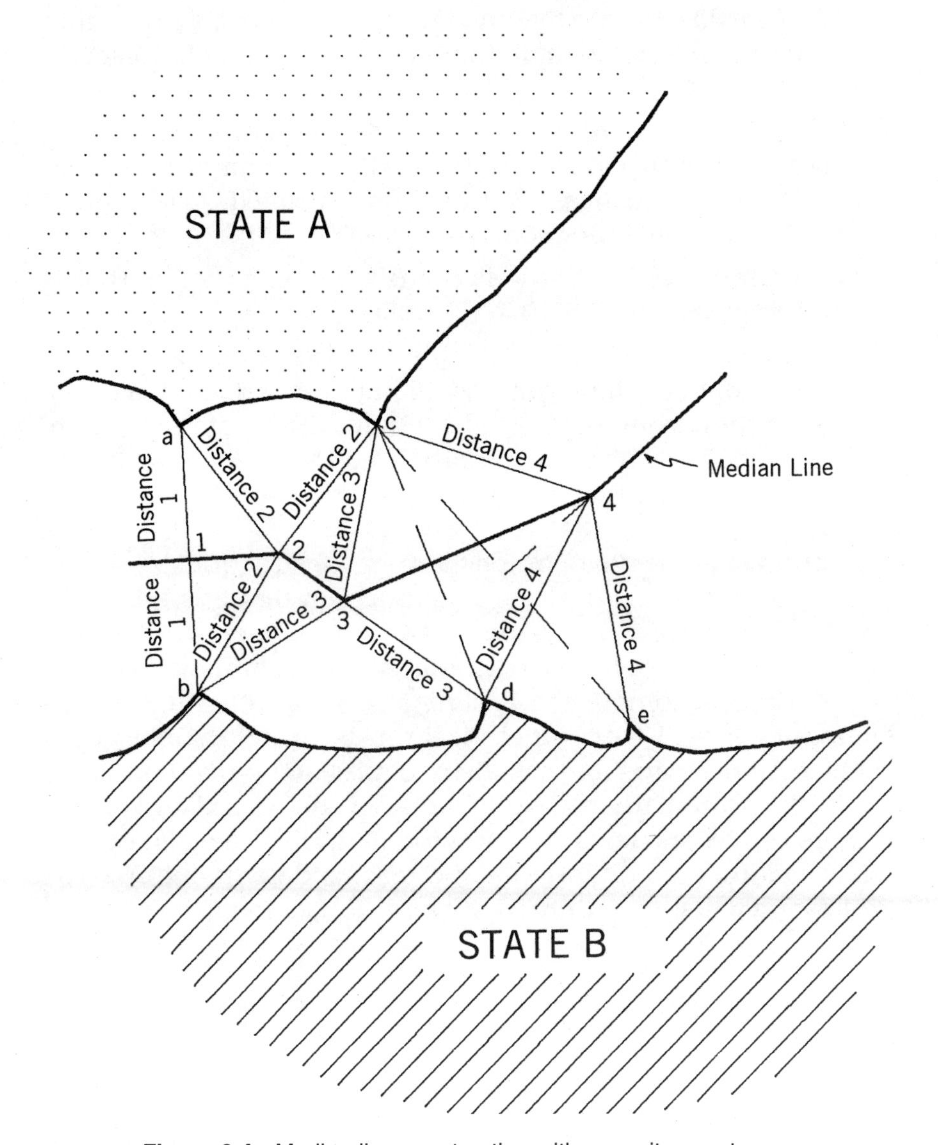

Figure 9.4 Median line construction with opposite coasts.

2. From Point 1, the median line runs from Point 1 (in either direction) along a line that is the perpendicular bisector of line ab.
3. The median line continues along the perpendicular bisector until Point 2 is found that is equidistant from Points a and b as well as from the next salient point (Point c) on either coast.
4. From Point 2, the median line runs along a line that is the perpendicular bisector of line cb.
5. As shown in Figure 9.4, this process is continued with the median line running along the perpendicular bisectors from the various equidistant points encountered. The line runs in both directions between the two opposing coasts.

Although somewhat complicated in construction, a median line is a technically desirable boundary in that it is based on precise measurements rather than more subjective determinations. Median line construction can be a cumbersome process when accomplished graphically. However, the process lends itself well to computerization using field surveyed or scaled coordinates for the salient shoreline points. For example, the coordinates of the initial point on a median line between opposite coasts (which is the midpoint between two salient shoreline points) may be easily determined by the following equations (Bureau of Land Management 1973):

$$N_m = \frac{N_a + N_b}{2} \tag{9.1}$$

$$E_m = \frac{E_a + E_b}{2} \tag{9.2}$$

Similarly, Cartesian coordinates for any of the inflection points on the median line may be computed by the following equations (Bureau of Land Management 1973):

$$N_m = \frac{H_a^2 - (E_b - E_a) - H_a^2 (E_c - E_a)}{2G} \tag{9.3}$$

$$E_m = \frac{H_a^2 - 2N_m - (N_b - N_a)}{2\,(E_b - E_a)} \tag{9.4}$$

where

a, b, c = subscripts designating Cartesian coordinates for the three salient shoreline points to which the inflection point is equidistant
m = subscript designating coordinates for the inflection point on the median line
N = Northing or y coordinate
E = Easting or x coordinate
G = $(N_c - N_a)\,(E_b - E_a) - (N_b - N_a)\,(E_c - E_a)$
H_a^2 = $E_b^2 + N_b^2 - E_a^2 - N_a^2$
H_b^2 = $E_c^2 + N_c^2 - E_a^2 - N_a^2$

For the initial point for a median line with adjacent coasts, the computation is slightly more cumbersome. In that case, the coordinates for the initial point may be established by distance-distance intersection. Equations for such a calculation are found in most elementary surveying texts.

There are certain geographic circumstances where an equidistant or median line may create inequities. These include areas where a small offshore island causes the line to curve substantially in one direction or where the coastline is generally convex. Where such configurations exist, other methods are warranted.

9.3 PROPORTIONALITY

One method often used to evaluate lateral and other common boundaries is the proportionality test. That technique compares the ratio of the areas allotted to two adjacent states by a potential boundary to the ratios of the lengths of the coastlines of the two states (Charney 1987). If the ratios are not similiar, the potential boundary may be inequitable. A variation of this test uses linear measurements instead of area. In a recent International Court of Justice case between the United States and Canada *(Gulf of Maine)*, the respective distances along the closing line for the Gulf was used. Linear proportionality has also been used in domestic cases involving the division of accretion, areas of exclusive riparian rights, and of the submerged lands in nonnavigable lakes.

9.4 CASE STUDY

To illustrate the principles of this chapter and problems involved in their application, a case study involving the division of offshore submerged lands between two U.S. coastal states is presented. The two states party to the boundary are Alabama and Mississippi. The configuration of the shoreline and the interstate land boundary is shown in Figure 9.5. The land boundary approaches the coast on a course slightly east of due south. The determination is complicated by the irregularity of the coastline near the termination of the upland boundary and by barrier islands located a few miles off the coast.

The upland boundary between the two states has an interesting history in itself. Mississippi was made a state from the western part of the Mississippi Territory in 1817. In the enabling act for the state *3 Stat. L. 472),* the southeastern and southern boundaries were described as starting at the northwest corner of the County of Washington and then running "due South to the Gulf of Mexico, Thence westwardly, including all of the islands within six leagues of the shore, to the most eastern junction of the Pearl River." Alabama was admitted into the Union in 1819 with a similiar description for the common boundary. A section of the enabling act for Alabama *(3 Stat. L. 490)* directed a survey of the common boundary with Mississippi and

> . . . if it should appear to said surveyors that so much of said line . . . will encroach on the counties of Wayne, Green or Jackson in said State of Mississippi, then the same shall be so altered as to run in a direct line from the north-west corner of Washington County to a point on the Gulf of Mexico, ten miles east of the mouth of the river Pascagola.

In 1820, the southern portion of the boundary was surveyed under the direction of General John Coffee and Major Thomas

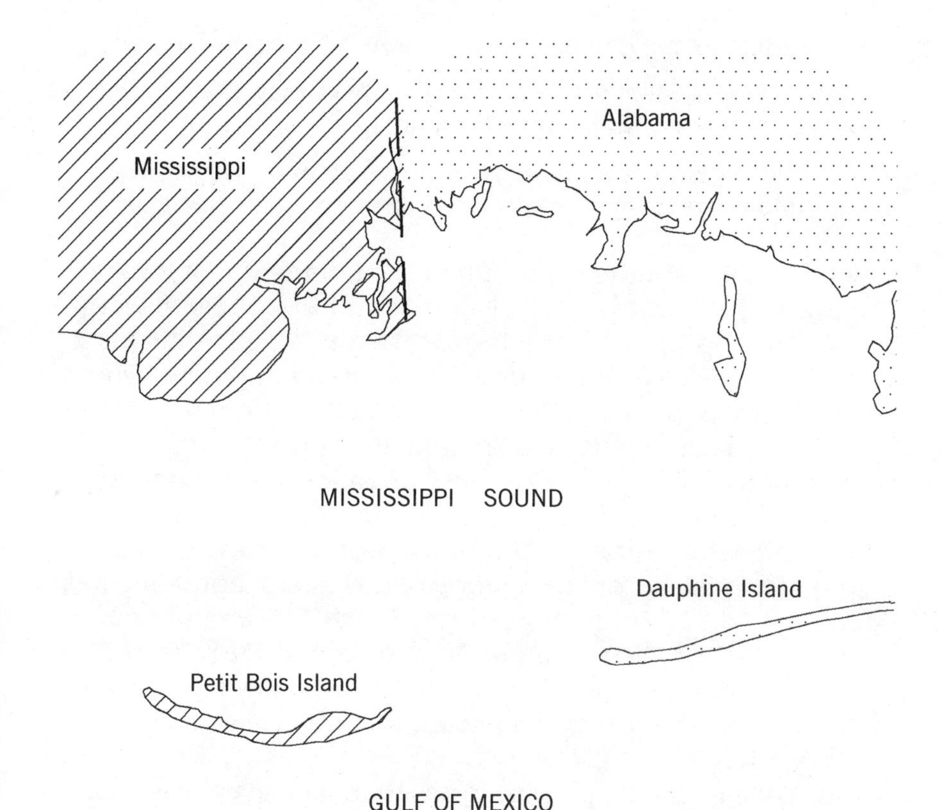

Figure 9.5 Upland boundary/shoreline configuration for Mississippi/Alabama boundary.

Freeman. After running a trial line due south, it appeared that the line did indeed encroach onto the border counties. Therefore, according to the field notes, the south terminus was moved easterly for "three miles, 57 chains and 40 links making 10 miles east from the meridian of the Pascagola." The final line was then run from that southern point, northerly to the original point of beginning at the northwest corner of Washington County. The true bearing of that part of the boundary was noted to be N 2° 08′ E (Van Zandt 1976).

It is interesting to note that the 1820 field notes described the land in the following terms:

> . . . horrible marshy, briary, bushy swamp—almost impassable
>
> This mile great deal worse in every respect than the former. Land unfit for everything, unfit for man, beast, fish or fowl.
>
> Thus far the land is apparently of no value and so thick with bush and briars that it is totally impassable except to surveyors.

Recently, the southern portion of the land boundary was resurveyed by the author, and it was found that these descriptions were appropriate. The terrain was extremely difficult to traverse, especially with modern-day survey equipment, which is considerably heavier than the compass and chain used in the original survey. Observations taken in this resurvey indicated that the line ran 2° 12′ E (astronomic), which compared favorably with the N 2° 08′ E as noted in the 1820 survey.

Regarding the lateral water boundary between the states, no record of an historic determination of this lateral boundary has been found. In 1976, after passage of the CEIP amendment to the Coastal Zone Management Act, correspondence from the Attorney General of Alabama to the National Oceanic and Atmospheric Administration indicated the following:

> The Alabama-Mississippi lateral seaward boundary has not been formally defined by statutory agreement but has traditionally been accepted as follows:
>
>> The seaward boundary between Alabama and Mississippi is traced by passing downward between Mobile County, Alabama, and Jackson County, Mississippi and upon reaching the Gulf of Mexico moves due south 00° 00′00″ to the seaward limit of each respective state.

Issue was taken with that definition by the State of Mississippi and negotiations were undertaken by the two states. An agreement was executed by the governors of the two states in 1978 for "purposes of determining which state is adjacent to particular outer continental shelf acreage for the purposes of the 1976 Amendment to the Coastal Zone Management Act of 1972 and for no other purposes." That agreement defined the boundary as follows:

> . . . the boundaries . . . extend from the point located at Latitude 30° 23′ 7″ and Longitude 88° 23′ 45″ to the point located at Latitude 30° 12′ 15″ and Longitude 88° 23′ 12″, the direction of said line being determined by the location of said two points. From said point last mentioned . . . the line . . . will extend South, 180° true azimuth. . . .

This description provides a lateral boundary approximating a straight-line extension of the land boundary to a point between Petit Bois and Dauphine Islands. At that point, the CEIP line runs due south.

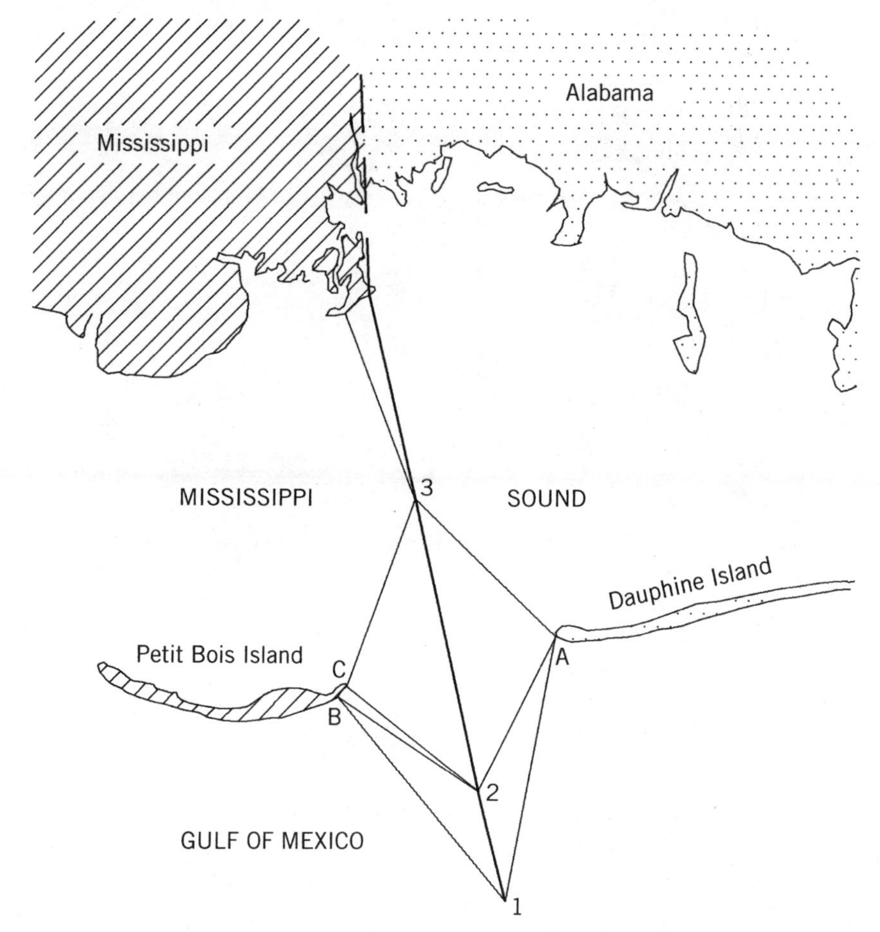

Figure 9.6 Median line construction for Mississippi/Alabama lateral boundary.

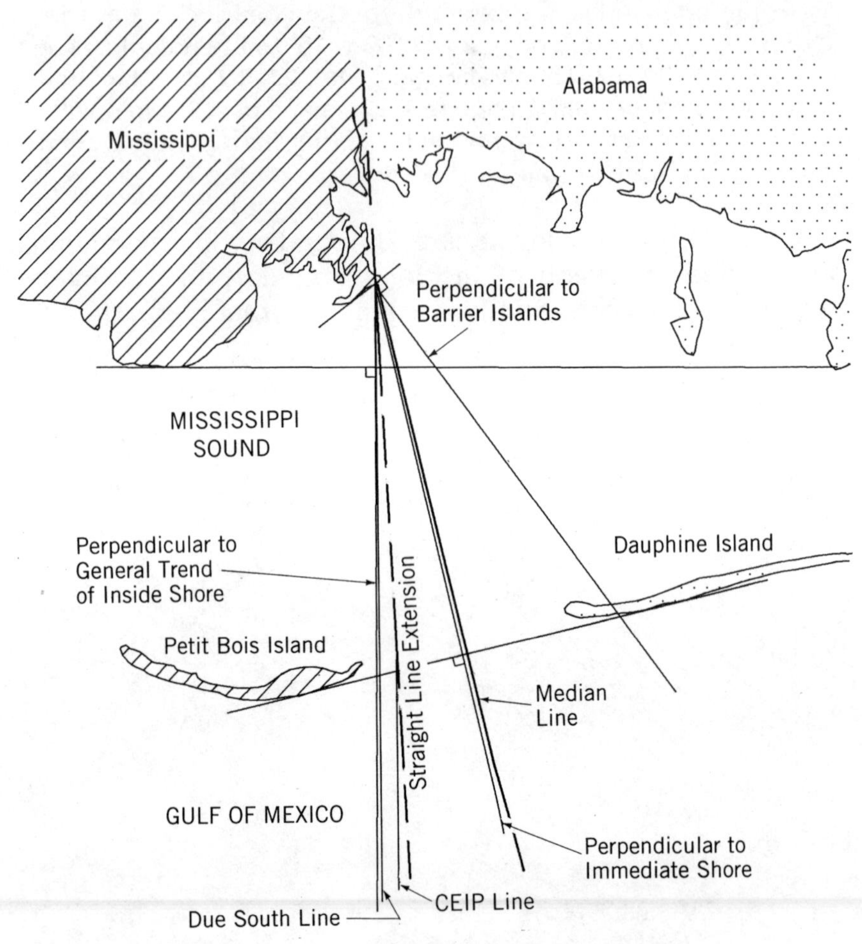

Figure 9.7 Alternatives for Mississippi/Alabama lateral boundary.

Recently, efforts have begun to negotiate a permanent lateral boundary between the two states to replace this special-purpose boundary. An excellent insight into the difficulties involved in developing a boundary agreement between these two states is obtained by examining the options for the boundary. To illustrate this, the following options for a lateral boundary in this case are examined:

1. Straight-line projection of land boundary
2. Perpendicular to shoreline
    (a) To the shoreline in the immediate area
    (b) To the general trend of the shoreline
    (c) To the trend of the shoreline of the barrier islands
3. Due south (cardinal direction)
4. Median line

Figure 9.6 shows the development of a median line for the area. The presence of the barrier islands complicates the construction, but a line may be drawn by careful application of the previously discussed procedures. Figure 9.7 shows the resulting median line compared with lines resulting from other options and with the CEIP line. These options provide a wide range of possible boundary lines.

Because negotiations are ongoing in this case as this text is being written, no comments will be made on the relative merits of the various options presented. Rather, it is left to the reader to weigh the options and determine the most equitable solution.

10

NATIONAL WATER BOUNDARIES

10.1 HISTORIC OVERVIEW

As mentioned in Chapter 8, the breadth of the coastal zone claimed by various coastal nations for their exclusive use has traditionally been 3 miles based on the informal "canon shot rule." Today, as various nations have begun exploiting the riches of the sea and sea floor, widely ranging claims have been made that vary considerably from the cannon shot rule. This issue has been frequently debated at international conferences on the law of the sea. At the 1930 Hague Conference of the International Law Commission and at the 1958 United Nations Conference on the Law of the Sea in Geneva, attempts were made to adopt a standard width zone without success. In 1982, a United Nations conference succeeded in doing so. That conference adopted an international law of the sea treaty giving coastal nations a territorial sea running out to 12 miles from shore with exclusive economic zones of 200 miles or to a maximum of 350 miles if the country's continental shelf extends out that far. This has resulted in more or less stabilizing these boundaries once again.

Reflecting claims by other coastal nations, the United States claim has also changed considerably during the last half century. Prior to the 1930s the United States claimed the 3-mile territorial sea and made no attempt to exert jurisdiction beyond that limit.

During the 1930s, in connection with enforcement of prohibition laws (rum runners had been able to easily escape interception with the 3-mile territorial jurisdiction), the jurisdiction of the United States for law enforcement was extended to the limits of a 12-mile "contiguous zone." It is noted that this jurisdictional extension was of an extraterritorial nature and did not extend the territorial sea ownership. Under Article 24 of the United Nations Convention on the Territorial Sea and Contiguous Zone, this jurisdiction is limited to that necessary to prevent infringement of customs, fiscal, immigration, or sanitary regulations within U.S. territory, or to punish such infringements (Colson 1982).

In 1945, President Truman further extended the claimed jurisdiction of the United States by a proclamation stating ". . . the Government of the United States regards the natural resources of the subsoil and seabed of the continental shelf beneath the high seas but contiguous to the coasts of the United States as appertaining to the United States, subject to its jurisdiction and control" *(Truman 1945; 43 USC Sec. 1332(1))*. The continental shelf is a gently sloping plain of land along the coasts of most islands and continents. It varies greatly in width, from a few miles to hundreds of miles. It is considered to end where the continental slope begins to drop more steeply to the ocean floor. The limit of the continental shelf and of this claim is currently defined as out to "a depth of 200 meters or, beyond that limit, to where the depth of the superacent water admits of the exploitation of the natural resources of the said area . . ." *(1958 Convention on the Continental Shelf, Article 1)*.

In 1977, following claims by other nations and failure of the United Nations Law of the Sea Conference to develop standards, the United States claimed exclusive fishing management jurisdiction to the 200-mile limit, regardless of whether this line went beyond the continental shelf *(16 USC Sec. 181 1)*. In 1982, as previously mentioned, a United Nations conference adopted a treaty establishing a 12-mile territorial sea. However, the United States did not become a party to this treaty reportedly due to disagreements over how deep sea mining should be administered and its profits shared. In 1988, however, the United States adopted an equivalent-width territorial sea. This occurred when President Ronald Reagan extended the U.S. territorial sea ". . . to the limits permitted by international law . . ." with a presidential proclamation (Reagan 1989). That document proclaimed as follows:

> The territorial sea of the United States henceforth extends to 12 nautical miles from the baseline of the United States determined in accordance with international law.

Therefore, the current posture of the United States is to claim ownership of a territorial sea of 12 nautical miles from its coastline; to claim exclusive jurisdiction over the natural resources to the extent of the continental shelf; and to claim exclusive fishing rights to 200 miles.

It is interesting that prior to the 1988 Reagan Proclamation, the United States claimed only a 3-mile territorial sea, whereas the states of Florida and Texas claimed a 3-league territorial sea in the Gulf of Mexico based on U.S. Supreme Court decisions. Therefore, the outer two-thirds of the state territorial sea was not opposable by the United States against other nations (Colson 1982). Presumably, these two states would have had to provide their own defense against foreign occupation within those areas.

10.2 BASELINES FOR NATIONAL BOUNDARIES

As mentioned in previous sections, offshore boundaries, such as the national water boundaries covered in this chapter, are usually defined as a line a certain number of miles seaward of the coastline. The coastline is in turn defined as the low water line depicted on charts of the area (which in the United States is currently the mean lower low water line). Obviously, when indentations in the coastline occur or when islands or other unusual coastline configurations are encountered, there are various interpretations that could be made. This would impact state/federal as well as international boundaries. Therefore, the United Nations Law of the Sea Conferences have developed rigid guidelines for baselines for various coastal configurations. These guidelines also have been adopted by case law for state/federal boundaries. Some of these guidelines are outlined below.

Rivers. Where a river flows directly into the sea, the baseline runs in a straight line across the river between headland points on the low tide line on either side of the mouth of the river.

Bays. The baseline at the mouth of bays, as discussed in Chapter 6, runs along the closing line drawn between the headlands of the bay. This applies for either a juridical or historic bay.

Low-Tide Elevations. A low-tide elevation is defined as "a naturally formed area of land which is surrounded by and above water at low tide but submerged at high tide" *(Article 3, 1958 United Nations Convention on the Law of the Sea).* Where low-tide elevations exist within the territorial sea, when measured from the mainland or an island, then the low water line for those elevations may be used as the baseline.

Islands. An island is defined as a "naturally formed area of land, surrounded by water, which is above water at high tide" *(Article 10, 1958 United Nations Convention on the Law of the Sea).* Islands have their own baseline from which to measure. As with the mainland, the baseline is the low water line around the island.

Harbors. "For the purpose of delimiting the territorial sea, the outermost permanent harbor works which form an integral part of the harbor system shall be regarded as forming part of the coast" *(Article 8, 1958 United Nations Convention of the Law of the Sea).*

Regional Straight Baselines. In certain situations, straight baselines may be used even though the actual shoreline is irregular. There are four such situations (Prescott 1985). First, where the coast is deeply indented and cut into, Article 7 of the Convention allows the baseline to be run straight between salient points. Second, where the coast is fringed with islands and, third, where there are unstable coasts that are likely to retreat, the same article allows straight baselines. The fourth situation involves areas that are archipelagos where Article 47 of the Convention allows the use of straight baselines.

10.3 TECHNIQUES FOR LOCATING NATIONAL BOUNDARIES

For the most part, techniques for locating our national boundaries are similar to those described for locating state/federal boundaries in Section 8.3. The principal difference is the distance that the boundary is measured seaward from the baseline. In some locations where the United States adjoins Canada and Mexico, or where the territorial sea of the United States overlaps those of other nations, then common boundaries must be determined. For these areas, the principles covered in Chapter 9 ("Lateral Water Boundaries of States") would apply except in any areas where the boundary is specifically described by treaty.

11

BOUNDARIES IN NONSOVEREIGN WATERS

11.1 INTRODUCTION

Previous sections of this text have dealt with boundaries associated with publicly owned or sovereign water bodies. Water bodies whose submerged lands are privately owned are also frequently used as boundaries because they offer a relatively permanent, easily recognized monument. Therefore, the following sections address such boundaries. Generally, nonsovereign waters are those nonnavigable waters that are not considered publicly owned under the public trust doctrine. However, this class of waters may also include navigable waters that have been conveyed out of the public trust at some point in time.

The boundary question involved in such waters is that of the extent of ownership of lands bordering these waters. In many respects, this problem is similiar to that of determining the limit of areas of exclusive riparian rights of owners of upland bordering sovereign waters (see Chapter 3). The difference is that these boundaries in nonsovereign waters delimit absolute ownership of the underlying soil as opposed to areas of certain exclusive rights. However, similiar methods of determination are used for both situations.

11.2 BOUNDARIES IN STREAMS

Perhaps the most elementary case of a nonsovereign water boundary is that involving a stream as the boundary between two parcels of land. In such a case where the deeds call "to the stream," the center of the stream is the boundary. There is general agreement on the part of most courts on this issue.

However, there are two complicating issues relating to stream boundaries in which there appears to be a divergence among court opinions. These areas deal with the questions of what constitutes the "center" of the stream and how the lateral boundaries between the upland boundaries and the center of the stream should be run.

Regarding the "center of the stream," one school of thought considers the boundary to be the "thread of the stream," which is defined as the line lying equidistant between the banks. For example, Bade (1940) states:

> The term "thread of the stream" means the geographic center of the stream at ordinary or medium stage of the water, disregarding slight and exceptional irregularities in the banks. It is fixed without regard to the main channel of the stream. . . . If the stream is made a boundary in a private conveyance, then the thread of the stream will be the stream boundary.

A number of court cases in several states support this position. An example *(Carlton v. Central and Southern Florida Flood Control District)* quotes *Blacks Law Dictionary* and defines the thread as ". . . a middle line; a line running through the middle of a stream."

The second approach to the center of the stream question holds the boundary to be the "thalweg" or deepest part of the channel. As with the former approach, a number of court cases also appear to support this opinion. One example *(Stubblefield v. Osborn)* that illustrates this position states:

> Upon principle, therefore, it would appear that the thread of a nonnavigable river is the line of water at its lowest stage. The thread or center of a channel, as the term is above employed, must be the line which would give to the landowners on either side access to the water, whatever its stage might be and particularly at its lowest stage.

On analysis, it appears that the latter approach is more equitable in that it allows access to the water even if the stream drops to an extremely low stage. Some suggest that whereas the deepest channel is the correct boundary for navigable, nonsovereign streams, the geographical center of the stream is the correct boundary for nonnavigable streams. On the surface, this may appear logical. However, this approach ignores that there are reasons for desiring access to water (such as recreational and consumptive uses) other than navigation. Denying access to one party to such a boundary appears inequitable.

Regarding the question of how lateral boundaries should be extended to the center of the stream, there is also more than one approach. The simplest method is that of straight-line extension of the upland boundary. Bade (1940) indicates that this would be the correct approach for unmeandered (not meandered in a Government Land Office survey) streams. However, if the stream is especially sinuous, which results in inequitable boundaries by the straight-line approach, Robillard and Bouman (1987) indicate that such partition lines should be drawn perpendicular to the "line of navigation." (Others suggest running perpendicular to the "center of stream.") A third approach, suggested by the *Manual of Instruction* (Bureau of Land Management 1973), apportions the length of a median line according to the length of a meander line of the shoreline of the adjacent riparian lots. This method is suggested when perpendicular lines do not result in equitable solutions.

The physical survey of stream water boundaries therefore involves first the establishment of the center of the stream. Depending on the circumstances and the prevailing law in the jurisdiction in which the stream is located, this could involve sounding the stream to find the deepest channel or, alternatively, finding the geographic center of the stream. The latter process results in the location of a more informal center "thread" or a median line, as described in Section 9.2. Once the center line is established, the upland side boundaries are extended to intersect this center line. Alternatively, for sinuous streams, a line is used that is perpendicular to the center line, or a line is used that connects the upland boundary line to points on the center line derived by proportioning it based on the length of shorelines of adjacent lots. The choice of methods should consider which solutions provide greater equity.

11.3 BOUNDARIES IN LAKES

Where a nonsovereign lake or pond forms a boundary line between two or more parcels, this can create a more complicated boundary problem than a stream boundary. This is due to the nonlinear shape of the lake and the lack of a main channel or current.

For lakes that are substantially round, the general rule is that a center point is selected at the geographical center of the lake. Partition lines are then run from the ends of the upland boundaries to the center point, forming "pie slices" *(Markusen v. Mortensen).* Arguably, a center point in the deepest portion in the lake could be chosen in lieu of the geographical center if the lake bottom has significant variations in depth. This would give equal access to the water if the lake drops to a low stage.

For long or irregular lakes, the stream formula is applied to the body of the lake and the round lake formula at the ends. This involves running a center line the length of the lake, along the geographical center line, to points that are the geographic center of arcs at the ends of the lake. At the ends, lateral lines converge at the end points. Along the length of the lake, lateral lines run to the center line at right angles to that line *(Rooney v. Stearns County Board).* As with streams, the center could be a median line or could be considered to be along the deepest portion of the lake. *Hardin v. Jordan* concisely states this approach:

> Where a lake is very long in comparison with its width, the method applied to rivers and streams would probably be the

most suitable for adjusting riparian rights in the lake bottom along its sides and the use of converging lines would only be required at its two ends.

Figure 11.1 shows the lake rules as obtained from plats contained in the opinions of two cited cases. Although these rules appear straightforward, application can be difficult. Such rules are far easier to state than to survey (Robillard and Bouman 1987).

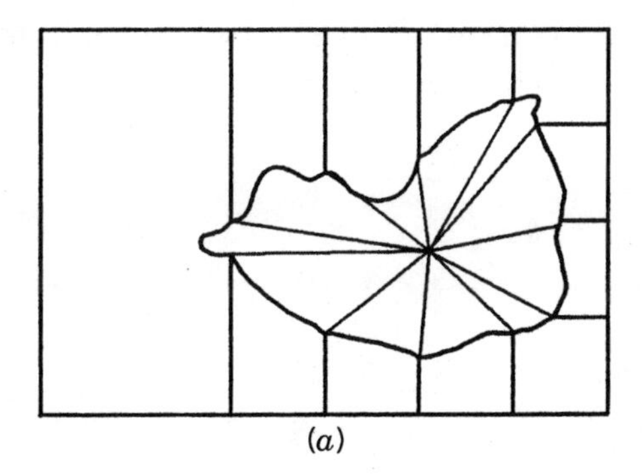

(*a*)

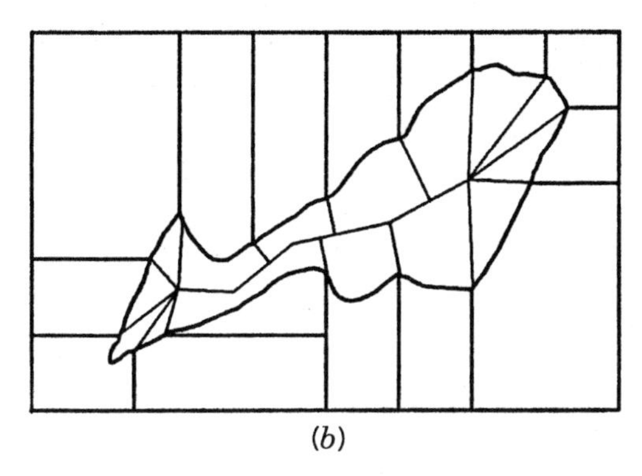

(*b*)

Figure 11.1 Method for proportioning lake bottoms: *(a)* round lakes (from *Markusen v. Mortensen*); *(b)* long lakes (from *Rooney v. Stearns County Board*).

11.4 CHANGES IN NONSOVEREIGN WATER BOUNDARIES

Generally, laws regarding changes in water boundaries due to physical changes in the water body are the same for private/private boundaries occurring in nonsovereign waters as they are for sovereign/private water boundaries. Therefore, they, too, are subject to change. For example, if a stream shifts position by the process of erosion on one side and accretion on the other, then the center of the stream—and with it the boundary—shifts with the stream. As with sovereign/private water boundaries, sudden changes due to avulsive activities generally do not result in a boundary shift. Also, in most states, a riparian owner may not benefit from shifts due to artificially caused accretion or erosion resulting from his or her actions.

APPENDIX

SPECIFICATIONS FOR MEAN HIGH WATER LINE SURVEYS

A.1 INTRODUCTION

The following are technical specifications for the survey of boundaries between public trust tidelands and submerged lands and adjacent uplands. These specifications apply to surveys of boundaries of all natural, tidally affected water bodies, where the boundary is considered to be the mean high water line. These specifications are for the purpose of locating the current mean high water line. Occasionally, due to avulsion or artificial shoreline alteration, the boundary may be deemed to be other than the current mean high water line. In such cases, procedures other than, or in addition to, those prescribed herein may have to be used. However, those procedures should be determined on a case-by-case basis.

In addition, unusual conditions encountered in some areas may require, or make desirable, deviation from these standards when surveying the current mean high water line. When this is the case, the justification for such a deviation will be noted on the survey plat.

A.2 DETERMINING LOCAL ELEVATION OF MEAN HIGH WATER

The first step in conducting a mean high water line survey is to determine the local elevation of mean high water. Due to a number of hydrographic conditions, this elevation may vary considerably from location to location. Therefore, it is important to use an elevation established in the immediate area of the proposed survey. The elevation will be determined as described in this section. For some land tracts, when a portion of the land borders one water body and another portion borders a restricted tributary to that water body, more than one method may have to be used.

A.2.1 Direct Use of Existing Tide Stations

A.2.1.1 If the water boundary being surveyed borders on a relatively unrestricted water body, is not near an inlet or an entrance to a subordinate water body, and is within a mile or so of an existing tide station located in the same water body, the elevations determined at that tide station may be used directly.

A.2.1.2 Locations of existing tide stations along with descriptions and elevations of the bench marks for each station may be obtained from an appropriate state agency or from the National Ocean Service. When using an existing tide station, at least two bench marks will be recovered and leveled between to check for mark disturbance, errors in published data, or mark misidentification.

A.2.2 Interpolating Between Tide Stations

A.2.2.1 If the water boundary being surveyed borders on the open coast and is not near an inlet, linear interpolation between adjacent tide stations on the open coast may be used.

A.2.2.2 Linear interpolation will be accomplished by determining the elevation of mean high water in relation to the National Geodetic Vertical Datum (NGVD, formerly known as sea-level datum) at a tide station on either side of the proposed survey. This will be done by leveling between the tidal bench marks and bench marks previously tied to NGVD. The elevation of mean high water at the proposed survey site will then be determined by proportioning the difference in mean high water elevation at the two stations, according to the distance between the two stations and the survey site.

A.2.3 Establishing Additional Short-Term Tide Stations

A.2.3.1 Where the criteria described for use of Sections A.2.1.1 and A.2.2.1 are not met, an additional tide station will be established. Depending on the configuration of the boundary of the tract being surveyed, more than one additional tide station may be required.

A.2.3.2 Additional tide stations will be established by simultaneous tidal observations between the new station and an established NOS tide station. With a densified network of existing tide stations, short-term observations are usually sufficient for establishing such stations. Such observations will be conducted for a minimum of three tidal cycles and should achieve a standard deviation of 0.10 foot or less, in the resulting elevation determinations. The resulting equivalent 19-year mean high water will be computed by one of the methods described in Section A.5.

A.2.3.3 For areas without a densified network of tide stations, it may be necessary to conduct a full month or more of simultaneous observations to achieve a satisfactory correlation. The length of observation depends on the location of the closest control station in relation to the survey site. When an additional tide station is established, the resulting elevation will be preserved by the establishment or level ties to one or more stable bench marks in the immediate vicinity of the survey.

A.2.4 Adjusting Elevation for Sea-Level Rise

A.2.4.1 Due to global sea-level changes as well as subsidence or upheaval of coastal lands over time, the current sea level may be different than that used in published data for tide stations. Therefore, a check should be made to estimate any such difference and a correction applied if the estimated difference exceeds 0.10 foot.

A.2.4.2 The correction factor will be estimated by obtaining a history of annual mean sea-level values for the primary station used in computing the tidal datum at the tide station being utilized. These data are available from the sources identified in Section A.2.1.2. The difference between mean sea level over the most recent 19-year epoch for which data are available and mean sea level over the 19-year epoch on which the datum was computed for publication will be used as the estimated correction factor.

A.3 MAPPING THE MEAN HIGH WATER LINE

Once the local elevation of mean high water is determined, as prescribed in Section A.2, the mean high water line will be located as specified in Section A.3.1 or A.3.2 and mapped as specified in Section A.3.3 or A.3.4.

A.3.1 Leveling

When using this method, the mean high water line will be located on the ground by leveling from an established tidal bench mark or from a tide staff on which the elevation of mean high water has been determined by one of the methods prescribed in Section A.2. This will be accomplished by assuming that mean high water is a contour in the immediate area and tracing that contour by conventional leveling methods.

A.3.2 Observing the Leading Edge of Water

When using this method, the mean high water line will be located on the ground by observing and staking the leading edge of an incoming tide at the exact time that the water level reaches mean high water. The time the water level reaches mean high water will be determined by observations on a graduated tide staff for which the elevation of mean high water has been determined by methods prescribed in Section A.2.

A.3.3 Horizontal Ties

After the line has been located on the ground as prescribed in Section A.3.1 or A.3.2, it will be tied by survey measurements to at least two survey monuments in the immediate vicinity of the survey.

A.3.4 Photo Interpolation

If only a graphic depiction of the location of the mean high water line is required, as opposed to an on-the-ground surveyed and staked line, the mean high water line may be mapped by photo interpolation. This will be accomplished by the location of infrequent ground truth points on the mean high water line using methodology prescribed in Sections A.3.1 and A.3.2 and then interpolating between such points using the tones, hues, and textures of aerial photography. If measurements are to be made on such photo maps, the photography will be scaled and rectified by surveyed ground control.

A.4 PLATTING REQUIREMENTS

The resulting plat of the mean high water line survey will contain the following information:

1. Tide station(s) used as control for the survey
2. Tidal bench marks used
3. 19-year epoch on which the tide station data were computed
4. Date of location of the mean high water line
5. Sufficient horizontal survey data to determine the relationship between inflection points on the mean high water line and local survey monumentation
6. If additional tide stations have been established, dates of simultaneous observations, method of computation used, and descriptions and tidal elevations for tidal bench marks established

7. If interpolation has been used, geodetic elevations of mean high water at the tide stations used and at the survey site

A.5 SECONDARY DATUM COMPUTATIONS

A.5.1 Standard Method

A.5.1.1 When the Standard Method is used, both high and low water must be observed at both the control and subordinate stations for each cycle of observation. The water level halfway between the observed high and low water at each station is considered to be the half-tide level and the difference between the observed high and low water is the tide range.

Equations to be used for the Standard Method are as follows:

$$MR_s = \frac{(MR_c)\,(R_s)}{R_c} \tag{A.2}$$

$$MTL_s = TL_s + MTL_c - TL_c \tag{A.1}$$

$$MHW_s = MTL_s + \frac{MR_s}{2} \tag{A.3}$$

where

MHW = 19-year mean high water
MTL = 19-year mean tide level
MR = 19-year mean tidal range
TL = mean tide level for observed tidal cycle(s)
R = mean range for observed tidal cycle(s)
s = subscript to denote subordinate station
c = subscript to denote control station

A.5.2 Amplitude Ratio Method

A.5.2.1 When it is desired to determine the elevation of mean high water where only the upper portion of the tidal cycle is observed at the subordinate station, the Amplitude Ratio

Method may be used. This situation may occur in intertidal zones. When this method is used, simultaneous tidal observations shall be taken for at least 2 hours before and after the high water peak for diurnal tides. One hour before and after are acceptable for semidiurnal tides. In addition, the low water extreme must be observed at the control station.

A.5.2.2 For computations using the Amplitude Ratio Method, an arbitrary time interval is selected to intersect the observed tide curves of the control and subordinate station. The same time interval must be used for tide curves at both stations. The vertical distance between the observed peak high water and the height on the tide curves, where they are intersected by the selected time interval, are used to determine range.

A.5.2.3 Equations to determine the range for the Amplitude Ratio Method are as follows:

$$R_s = \frac{R_c A_s}{A_c} \tag{A.4}$$

$$\mathrm{MR}_s = \frac{(\mathrm{MR}_c)\ (A_s)}{A_c} \tag{A.5}$$

where

A = observed vertical distance between high water and time interval line
R = tidal range for observed cycle
MR = 19-year mean range
s = subscript to denote subordinate station
c = subscript to denote control station

Mean tide level and mean high water at the subordinate station can then be determined by Equations (A.2) and (A.3) of the Standard Method.

A.5.3 Height Difference Method

A.5.3.1 The Height Difference Method may be used when only the upper portion of the tidal cycle is observed at the subordinate station if the difference in range between the control and subordinate stations is less than 10 percent.

A.5.3.2 The equation for the Height Difference Method is as follows:

$$MHW_S = HW_S + MHW_C - HW_C \tag{A.6}$$

where

HW = mean high water of observed cycle(s)
MHW = 19-year mean high water
S = subscript to denote subordinate station
C = subscript to denote control station

REFERENCES

Bade, Edward S. (1940). "Title, Points, and Lines in Lakes and Streams." *Minnesota Law Review*, 24: 305.

Beazley, R. B. (1978). *Maritime Limits and Baselines*, 2nd ed., Special Pub. No. 2. London: The Hydrographic Society.

Bodnar, A. Nicholas, Jr. (1977). *User's Guide for the Establishment of Tidal Bench Marks and Leveling Requirements for Tide Stations.* U.S. Government Printing Office, Washington, D.C.: National Ocean Survey.

Bouman, Lane J. (1977). "The Meandering Process in the Survey of the Public Lands of the United States," in *Proceedings of The Water Boundary Workshop*, California Land Surveyors Association, pp. 231–235.

Bowditch, Nathaniel (1962). *American Practical Navigator*, Pub. No. 9. U.S. Government Printing Office, Washington, D.C.: U.S. Hydrographic Office.

Briscoe, John (1984). "Federal-State Offshore Boundary Disputes: The State Perspective," in *Proceedings of the 18th Annual Law of the Sea Institute Conference.* San Francisco, pp. 1–78.

Bureau of Land Management (1973). *Manual of Instruction for the Survey of the Public Lands of the United States.* U.S. Government Printing Office, Washington, D.C.

Charney, Jonathan I. (1989). "The Delimitation of Ocean Boundaries," in *Rights to Oceanic Resources*, Boston: Martinus Nijhoff Publishers, pp. 25–49.

Christie, Donna R. (1979). "Coastal Energy Impact Program Boundaries on the Atlantic Coast." *Virginia Journal of International Law*, 19: (4) 841–881.

Coast and Geodetic Survey (1965). *Manual of Tide Observations*, Publication 30-1. U.S. Government Printing Office, Washington, D.C.

Cole, George M. (1977). "Tidal Boundary Surveying." *Technical Papers,* American Congress on Surveying and Mapping, (Spring).

——— (1979). "Non-Tidal Water Boundaries." *Florida Bar Real Property, Probate and Trust Law Newsletter,* (January).

——— (1981). "Proposed New Method for Determining Tidal Elevations in Inter-tidal Zones." *Technical Papers,* American Congress on Surveying and Mapping, (Spring): 356–366.

——— (1982). "Where Oil, Water, Surveying and Photogrammetry Mix." *Technical Papers,* American Congress on Surveying and Mapping, (Spring): 319 323.

——— (1988). "Use of Hydrology for Determining Ordinary High Water in Non-Tidal Waters." *Technical Papers,* American Congress on Surveying and Mapping, (Spring): 34–43.

——— (1990). "The Significance of the Meandering of Water Bodies in Public Land Surveys." *Surveying and Land Information Systems,* American Congress on Survveying and Mapping, 50: (3): 233–236.

——— (1991). "Tidal Water Boundaries." *Stetson Law Review,* 20 (1 and 2): 165–176.

——— (1996). "Estuarine Tidal Range Variation and Hydrographic Causes." M.S. thesis, Florida State University, Tallahassee, FL.

Cole, George M., and Robert J. Dearsing (1989). "Tidal Data Acquisition—Texas Style," in *Proceedings of the Coastal Zone '89.*

Cole, George M., and F. Michael Speed (1991). "Use of Constituent Analysis for Estimation of Tidal Data by Simultaneous Short-Term Observations." *Technical Papers,* American Congress on Surveying and Mapping, (Spring): 44–53.

Cole, George M., F. Michael Speed, and James C. Fugate (1989). "Use of Regression Analysis for Estimating Tidal Data," in *Proceedings of the Marine Technology Society.*

Colson, David A. (1982). U.S. Department of State, personal communication.

Davis, J. H. (1972). *Establishment of Mean High Water Lines in Florida's Lakes,* Pub. No. 24. Gainesville, FL, University of Florida, Florida Water Resources Research Center.

Defant, Albert (1958). *Ebb and Flow.* Ann Arbor, MI: University of Michigan Press.

Duxbury, Alyn C., and B. Alison (1989). *An Introduction to the World's Oceans.* Dubuque, IA: William C. Brown.

C. T. Foster (1959). *Annotation on Apportionment and Division of*

Area of River as Between Riparian Tracts Fronting on Same Bank in Absence of Agreement or Specification. 65 American Law Reports 2nd 143.

General Land Office (1902). *Manual of Surveying Instructions for the Survey of the Public Lands of the United States and Private Lands Claims.* Washington, D.C.: U.S. Government Printing Office.

Gentry, Daniel (1989). Jones Wood & Gentry, Orlando, Florida, personal communication.

Hale, M. (1666). *De Jure Maris.* Reprinted in S. Moore, *A History of the Foreshore and the Law Relating Thereto,* 3rd ed. (1888)

Hanna, John (1951). "The Submerged Lands Cases." *Baylor Law Review,* 3: 201–240.

Hawes, J. H. (1977). *Manual of United States Surveying.* Reprinted by Carben Surveying Reprints; originally published by J. B. Lippincott, 1868.

Hicks, Stacy D., Henry A. Debaugh, and Leonard E. Hickman, Jr. (1983). *Sea Level Variations for the United States 1855–1980.* Washington, D.C.: U.S. Department of Commerce, NOAA, NOS.

Hicks, Stacy D., and Leonard E. Hickman, Jr. (1988). "United States Sea Level Variations Through 1986." *Shore and Beach,* American Shore and Beach Preservation, (July): 3–7.

Hodgson, Robert D., and Lewis M. Alexander (1972). *Towards an Objective Analysis of Special Circumstances,* Occasional Paper No. 13. Kingston, RI: University of Rhode Island, Law of the Sea Institute.

Kapoor, D. C., and Adam J. Kerr (1986). *A Guide to Maritime Boundary Delimitation.* Toronto: Carswell.

Knochenmus, D. D. (1967). *Shoreline Features as Indicators of High Lake Levels,* Professional Paper No. 575-C. Washington, D.C.: U.S. Geological Survey.

Krumbein, William C. (1963). *Stratigraphy and Sedimentation.* San Francisco, CA: W. H. Freeman.

Lyles, Steve D., Leonard E. Hickman, Jr., and Henry A. Debaugh (1988). *Sea Level Variations for the United States 1855–1986.* Washington, D.C.: U.S. Department of Commerce, NOAA, NOS.

Maloney, Frank E. (1978). "The Ordinary High Water Mark: Attempts at Settling an Unsettled Boundary Line." *University of Wyoming Land and Water Law Review,* 13(2): 1–30.

Maloney, Frank E., and Richard C. Ausness (1974). "The Use and Legal Significance of the Mean High Water Line in Coastal Boundary Mapping." *The North Carolina Law Review*. (December): 185–273.

Marmer, H. A. (1951). *Tidal Datum Planes,* rev. ed., Special Publication No. 135. Washington, D.C.: U.S. Coast and Geodetic Survey.

National Ocean Survey (1975). *Tide and Current Glossary.* Washington, D.C.: U.S. Department of Commerce.

——— (1976). *Our Restless Tide.* Washington, D.C.: U.S. Department of Commerce.

——— (1980). *The National Tidal Datum Convention of 1980.* Washington, D.C.: U.S. Department of Commerce.

Pearcy, G. Etzel (1959). "Geographical Aspects of the Law of the Sea." *Annals of the Association of American Geographers,* (March): 1–19.

Prescott, J. R. V. (1985). *The Maritime Political Boundaries of the World.* London: Methuen.

Reagan, Ronald (1989). *Territorial Sea of the United States of America,* Proclamation 5928. *Federal Register,* 54 (W05).

Redfield, Alfred C. (1950). "The Analysis of Tidal Phenomena in Narrow Embayments." *Papers in Physical Oceanography and Meteorology,* 11(4): 1–37.

Robillard, Walter G., and Lane J. Bouman . (1987). *Clark on Surveying and Boundaries,* 5th ed. Charlottesville, VA: Michie Publishers.

Sandars, Thomas C. (1874). *The Institutes of Justinian.* London: Longmans, Green & Co.

Schureman, Paul (1958). *Manual of Harmonic Analysis and Prediction of Tides,* U.S. Government Printing Office, Washington, D.C.

Shalowitz, Aaron L. (1962). *Shore and Sea Boundaries,* Pub. No. 10-1. Washington, D.C.: U.S. Coast and Geodetic Survey.

Stiles, Arthur A. (1952). "The Gradient Boundary." *The Texas Law Review,* 30(3): 305–322.

Sullivan, D. E. (1982). "Annotation on Allocation of Water Space Among Lakefront Owners in Absence of Agreement or Specification." 14 *American Law Reports* 4th 1028.

Swanson, L. W. (1982). Captain, USC&GS, Retired, personal communication.

Swanson, R. Lawrence (1974). *Variability of Tidal Datums and Accuracy in Determining Daturns from Short Series of Observations,* Technical Report No. 64. National Ocean Survey. U.S. Government Printing Office, Washington, D.C.

Thornbury, William D. (1954). *Principals of Geomorphology.* New York: John Wiley.

Truman, H. S. (1945). *The Continental Shelf,* Proclamation No. 2667 (59 Stat. 884).

Tucker, K. (1983). Deputy Attorney General, State of Florida, personal communication.

U.S. Army Corps of Engineers (1977). *Shore Protection Manual.* Coastal Engineering Research Center.

U.S. Coast Survey (1898). *Annual Report.* Washington, D.C.: U.S. Department of Commerce.

U.S. Naval Facilities Engineering Command (1982). *Coastal Protection,* Design Manual 26.2. U.S. Government Printing Office, Washington, D.C.

Van Zandt, Franklin K. (1976). *Boundaries of the United States and the Several States,* Professional Paper 909. Washington, D.C.: U.S. Geological Survey.

Williams, Jerome (1962). *Oceanography.* Boston: Little, Brown.

Wood, Fergus J. (1976). *The Strategic Role of Perigean Spring Tides in Nautical History and North American Coastal Flooding.* U.S. Government Printing Office, Washington, D.C.: National Ocean Survey.

Zetler, Bernard D. (1959). "Tidal Characteristics from Harmonic Constants." *Proceedings of the American Society of Civil Engineers,* 85 (HY 12).

——— (1981). "Methods of Estimating Mean High Water from Partial Data Curves." *Technical Papers,* American Congress on Surveying and Mapping, (Spring): 367–381.

LAW CASE CITATIONS

Aborn v. Smith, 12 R.I. 370, 1879.

American Steel & Wire Co. V. Cleveland Electric Illuminating Co., 16 Ohio LR 250, 1909.

Application of Central Nebraska Public Power & Irrigation District, 295 NW 386, Mich. 1940.

Attorney General v. Chambers, 17 Eng. Rul. Cas. 555, 1854.

Baker v. State ex rel Jones, Fla., 87 So. 2d. 497, 1956.

Baltimore v. Baltimore & PSB Co., 65 A 353, Md., 1906.

Barney v. Keokuk, 94 U.S. 324, 1876.

Bay City Gaslight Co. v. Industrial Works, 28 Mich. 182, 1873.

Bliss v. Kinsey, Fla., 233 So. 2d. 191, 1970.

Blodgett & Davis Lumber co. v. Peters, 49 NW 917, 1891.

Board of Trustees v. Madeira Beach Nominee Inc., Fla., 272 So. 2d. 209, 1973.

Board of Trustees v. Sand Key Associates Ltd., 512 So. 2d. 209, Fla., 2d. DCA, 1973.

Board of Trustees v. Wakulla Silver Springs Co., Fla., 362 So. 2d. 706, 1978.

Bond v. Wool, 12 SE 281, N.C., 1890.

Borax Consolidated Ltd. v. City of Los Angeles, 296 U.S. 10, 1935.

Borough of Ford City v. United States, 345 F. 2d. 645, 1965.

Bradley v. McPherson, N.J., 59 A 105 1904.

Bright et al. v. Mauro et al., Cause No. 93-05265, Travis Co., District Court, 200th Judicial District, Tex., 1996.

Bucki v. Cone, Fla., 6 So. 160, 1889.

Calkins v. Hart, 11 3 NE 785, N.Y., 1916.

Campau Realty Co. v. Detroit, Mich., 127 NW 365, 1910.

Carlton v. Central and Southern Florida Flood Control District, Fla., 181 So. 2d. 656, 1966.

Cinque Bambini Partnership et al. v. State of Mississippi et al., Miss., 491 So. 2d. 508, 1986.

City of Tarpon Springs v. Smith, Fla., 88 So. 25, 1912.

Clark v. Campau, 19 Mich. 325, 1869.

Clement v. Watson, Fla., 58 So. 25, 1912.

Columbia Land Co. v. Van Dusen Investment Co., 19 P 469, Or., 1907.

Connerly v. Perdido Key, Inc., Fla., 270 So. 2d. 390, 1972.

Cordovana v. Vipond, Va, 94 SE 2d. 295, 1956.

Daniel Ball (The), 77 U.S. 557, 1870.

Deerfield v. Arms, 28 Am Dec 27, Mass., 1835.

Delaware L&WR Co. v. Hunnan, 37 NJL 276, 1875.

Dow v. Electric Company, N.H., 45 A 350, 1899.

Driesbach v. Lynch, 234 P 2d 446, Id., 1951.

Farris v. Bentley, 124 NW 1003, Wisc., 1910.

Fraiser's Million Dollar Pier Co. v. Ocean Park Pier Co., Ca., 197 P 328, C. 1921.

Goose Creek Hunting Club v. United States, 518 F 2d. 579, 1975.

Groner v. Foster, 27 SE 493, Va., 1897.

Gulf of Maine (Canada/United States), LC.J. 18, 83-84, 1982.

Hanson v. Rice, 92 NW 982, Minn., 1903.

Hardin v. Jordan, 140 U.S. 371, 1891.

Harrison v. Fite, 148 F 781, 1906.

Hayes v. Bowman, Fla., 91 So. 2d. 795, 1957.

Heard v. State, Tex., 204 S.W. 2d. 344, 1947.

Hefferline v. Langkow, 552 P 2d. 1079, Wa., 1976.

Hilleary v. Meyer, Id., 430 P 2d. 65, Id., 1975.

Howard v. Ingersoll, 54 U.S. 381, 427, 1851.

Johnson v. McCowen, Fla., 348 So. 2d. 357, 1977.

Kelly's Creek and N.W.R. Co. v. United States, 100 Ct. C1 396, 1943.

Knight v. Wilder, 48 Am Dec 660, Mass., 1848.

Korterud v. Darterud, 195 NW 972, S.D., 1923.

Lambert's Point Co. v. Norfolk & W.R. Co., 74 SE 156, Va., 1912.
Lattig v. Scott, 107 P 47, Id., 1910.
Lopez v. Smith, 109 So. 2d. 176, Fla. 2d. DCA, 1959.
Ludwig v. Overly, 19 Ohio CC 709, 1895.
Luttes v. State, 324 SW 2d. 167, 1958.

Manchester v. Point St. Iron Works, 13 R.I. 355, 1881.
Manufacturers Land & Improvement Co. v. Board of Commerce & Navigation, 121 A 337, N.J., 1923.
Markusen v. Mortensen, Minn., 116 NW, 1021.
Martin v. Busch, Fla., 112 So. 274, 1947.
Martin v. Waddell, 41 U.S. (16 Pet.) 367, 1842.
McCamon v. Stagg, 43 U.S. 367, 1896.
McCullough v. Wall, 53 Am Dec 715, S.C., 1850.
McDowell v. Trustees of the Internal Improvement Trust Fund of State of Florida, Fla., 90 So. 2d. 715, 1956.
Menasha Wooden Ware Co. v. Lawson, 36 NW 412, Wisc., 1888.
Mix v. Trice, 298 NYS 441, 1937.
Montgomery v. Shaver, 66 P 923, Or., 1901.
Motl v. Boyd, Tex., 286 W.W. 458, 1926.
Municipality No.2 v. Municipality No. 1, 17 LA 573, 1841.

Northern Pine Land Co. v. Bigelow, 54 NW 496, Wisc., 1893.

Odom v. Deltona, Fla., 341 So. 2d. 977, 1976.
Oklahoma v. Texas, 260 U.S. 606, 261 U.S. 340, 1923.
Oneal v. Rollinson, 192 SE 688, N.C., 1937.

People ex rel Cornwall v. Woodruff, 51 NYS 515, 1898.
Phillips Petroleum Co. v. Mississippi, 484 U.S. 469, 1988.
Pollard's Lessee v. Hagan, 3 How 212, 44 U.S., 1845.

Randall v. Ganz, P 2d. 65, Id., 1975.
Rector v. United States., 20 F 2d. 845, 1927.
Rice v. Standard Products Co., 99 SE 2d. 529, Va., 1957.
Rooney v. Stearns County Board, Minn., 153 NW 858, 1915.

San Francisco Savings Union v. Irwin, 28 F 708, 1886.
Scheifert v. Briegal, 96 NW 44, Minn., 1903.

Seattle Factory Sites Co. v. Saulsberry, 229 P 10, Wa., 1924.

Shedd v. American Maise Products Co., 108 NE 610, Ind., 1915.

South Shore Lumber Co. v. C. C. Thompson Lumber Co., 94 F 7738, Wisc., 1899.

State ex rel O'Conner v. Sorenson, Okla., 198 P 2d. 402.

State of Florida et al v. Florida National Properties, Fla., 338 So. 2d. 13, 1976.

Stuart v. Greanyea, 11, 7 NW 65 5, Mich., 1908.

Stubblefield v. Osborn, Neb., 31 NW 2d. 547, 1948.

Superior v. Northwestern Fuel Co., 161 NW 9, Wisc., 1917.

Tabor v. Hall, 51 A 432, R.I., 1902.

Thomas v. Ashland, 100, NW 993, Wa., 1904.

Tilden v. Smith, Fla., 113 So. 708, 1927.

Trustees of the Internal Improvement Trust Fund v. Ocean Hotels, Inc., 40 Fla. Supp. 26, 1974.

Trustees v. Wetstone, Fla., 222 So. 2d. 10, 1969.

Ulbright v. Basington, 119 P 292, Id., 1911.

U.S. v. 2,899.17 Acres of Land Etc., 269 F Supp. 903, 1967.

U.S. v. Alaska, 422 U.S. 184, 1975.

U.S. v. California, 332 U.S. 19, 1947; 432 U.S. 40, 1977.

U.S. v. Florida, 425 U.S. 791, 1976.

U.S. v. Joder Cameron, 466 Fed. Supp. 1099, 1979.

U.S. v. Louisiana, 339 U.S. 699, 1950; 394 U.S. 1, 10, 1969.

U.S. v. Louisiana, Texas, Mississippi, Alabama and Florida, 3 63 U.S. I. 212, 1960.

U.S. v. Maine, 470 U.S. 515, 1975.

U.S. v. Parker, No. 75-34, N.D. Fla., 1976.

U.S. v. Ruggles, 5 Blatch 35, 1861.

U.S. v. Texas, 339 U.S. 707, 1950.

Waverly Waterfront & Improvement Co. v. White, 33 SE 534, Va., 1899.

Wood v. Appal, 63 Pa. 210, 1869.

INDEX